1

COLLOQUIA

Yuri F. Bilu
A2X, Université Bordeaux I
351, cours de la Libération 33
405 Talence cedex, France

Corrado De Concini
Dipartimento di Matematica
Università di Roma "La Sapienza"
Piazzale Aldo Moro 5
00185 Roma, Italia

Simon Gindikin
Department of Math., Hill Center
Rutgers University
110 Frelinghysen Road
Piscataway, NJ 08854-8019

Dorian Goldfeld
Columbia University
Department of Mathematics
New York, NY 10027, USA

David Masser
Mathematisches Institut
Universität Basel
Rheinsprung 21
4051 Basel, Switzerland

Zeév Rudnick
School of Mathematical Sciences
Schrieber Building, Room 316
Tel-Aviv University
Tel-Aviv 69978, Israel

Lucien Szpiro
The Graduate Center
The City University of New York
365 Fifth Avenue
New York, NY 10016-4309, USA

Thomas J. Tucker
Department of Mathematics
University of Georgia
Athens, GA 30602-7403, USA

Colloquium De Giorgi 2006

Colloquium
De Giorgi 2006

edited by
Umberto Zannier

EDIZIONI
DELLA
NORMALE

ISBN: 978-88-7642-212-6

Colloquium
De Giorgi 2006

edited by
Umberto Zannier

EDIZIONI
DELLA
NORMALE

ISBN: 978-88-7642-212-6

Contents

Preface

Since 2001 the Scuola Normale Superiore di Pisa, connecting with an initiative during the years 1996, 1997 and 1998, has organized the "Colloquio De Giorgi", a series of colloquium talks named after Ennio De Giorgi, the eminent analyst who was a Professor at the Scuola from 1959 until his death in 1996.

The Colloquio takes place once a month. It is addressed to a general mathematical audience, and especially meant to attract graduate students and advanced undergraduate students. The lectures are intended to be not too technical, in fields of wide interest. They must provide an overview of the general topic, possibly in a historical perspective, together with a description of more recent progress.

The idea of collecting sistematically the materials from these lectures and publishing them in annual volumes came out recently, as a recognition of their intrinsic mathematical interest, and also with the aim of preserving memory of these events.

For this purpose, the invited speakers are now asked to contribute with a written exposition of their talk, in the form of a short survey or extended abstract.

This is the first volume in a collection which we hope will continue on a yearly basis. This first volume also contains a complete list of the conferences held in the "Colloquio De Giorgi" in the past years.

Conferences held in 2001

Paul Gauduchon
Weakly self-dual Kähler surfaces

Tristan Rivière
Topological singularities for maps between manifolds

Frédéric Hélein
Integrable systems in differential geometry and Hamiltonian stationary Lagrangian surfaces

Jean-Pierre Demailly
Numerical characterization of the Kähler cone of a compact Kähler manifold

Elias Stein
Discrete analogues in harmonic analysis

John N. Mather
Differentiability of the stable norm in dimension less than or equal to three and of the minimal average action in dimension less than or equal to two

Guy David
About global Mumford-Shah minimizers

Jacob Palis
A global view of dynamics

Alexander Nagel
Fundamental solutions for the Kohn-Laplacian

Alan Huckleberry
Incidence geometry and function theory

Conferences held in 2002

Michael Cowling
Generalizzazioni di mappe conformi

Felix Otto
The geometry of dissipative evolution equations

Curtis McMullen
Dynamics on complex surfaces

Nicolai Krylov
Some old and new relations between partial differential equations, stochastic partial differential equations, and fine properties of the Wiener process

Tobias H. Colding
Disks that are double spiral staircases

Cédric Villani
When statistical mechanics meets regularity theory: qualitative properties of the Boltzmann equation with long-range interactions

Conferences held in 2003

John Toland
Bernoulli free boundary problems - progress and open questions

Jean-Michel Morel
The axiomatic method in perception theory and image analysis

Jacques Faraut
Random matrices and infinite dimensional harmonic analysis

Albert Fathi
C^1 *subsolutions of Hamilton-Jacobi Equation*

Hakan Eliasson
Quasi-periodic Schrödinger operators - spectral theory and dynamics

Yakov Pesin
Is chaotic behavior typical among dynamical systems?

Don B. Zagier
Modular forms and their periods

David Elworthy
Functions of finite energy in finite and infinite dimensions

Conferences held in 2004

Jean-Christophe Yoccoz
Hyperbolicity for products of 2×2 *matrices*

Giovanni Jona-Lasinio
Probabilità e meccanica statistica

John H. Hubbard
Thurston's theorem on combinatorics of rational functions and its generalization to exponentials

Marcelo Viana
Equilibrium states

Boris Rozovsky
Stochastic Navier - Stokes equations for turbulent flows

Marc Rosso *Braids and shuffles*

Michael Christ
The d-bar Neumann problem, magnetic Schrödinger operators, and the Aharonov-Böhm phenomenon

Conferences held in 2005

Louis Nirenberg
One thing leads to another

Viviane Baladi
Dynamical zeta functions and anisotropic Sobolev and Hölder spaces
Giorgio Velo
Scattering non lineare

Gerd Faltings
Diophantine equations

Martin Nowak
Evolution of cooperation

Peter Swinnerton-Dyer
Counting rational points: Manin's conjecture

François Golse
The Navier-Stokes limit of the Boltzmann equation

Joseph J. Kohn
Existence and hypoellipticity with loss of derivatives

Dorian Goldfeld
On Gauss' class number problem

Diophantine equations with separated variables

Yuri F. Bilu

Let $f(x)$ and $g(x)$ be polynomials with rational coefficients. One may ask whether the sets $f(\mathbb{Z})$ and $g(\mathbb{Z})$ have finitely or infinitely many elements in common. In other words, one asks whether the equation

$$f(x) = g(y) \tag{1.1}$$

have finitely or infinitely many solutions in $x, y \in \mathbb{Z}$.

Due to the classical theorem of Siegel [32], the finiteness problem for (1.1), and even a for a more general equation $F(x, y)=0$ with $F(x, y) \in \mathbb{Z}[x, y]$, is decidable.[1] The theorem of Siegel asserts that the absolutely irreducible equation $F(x, y) = 0$ has finitely many solutions in integers unless the plane curve defined by the equation is of genus 0 and has at most two points at infinity. Therefore, to solve the finiteness problem for the general equation $F(x, y) = 0$, one has to:

- decompose the polynomial $F(x, y)$ into $\mathbb{Q}$-irreducible factors;
- for those factors which are not $\overline{\mathbb{Q}}$-reducible, determine the genus **g** and the number d of points at infinity of the corresponding plane curve;
- for the factors with $\mathbf{g} = 0$ and $d \le 2$ determine whether the corresponding equation has finitely or infinitely many integral solutions (see [4, Section 1]).

Though this procedure completely solves the problem when the polynomials $f(x)$ and $g(x)$ are given numerically, it is not very helpful when they depend on unknown parameters: one wants to have something more explicit.

---

[1] Mention in passing that the decidability of the *existence* problem (that is, whether or not $F(x, y) = 0$ has at least one solution) is still an open question. The answer, which is believed to be positive, is known only in particular cases.

In this talk I will speak on the very explicit finiteness criterion for the equation (1.1), obtained jointly with R. F. Tichy [9]. It turns out to be more convenient to study a slightly more general problem: when does (1.1) have infinitely many rational solutions with a *bounded denominator*? We say that the equation $F(x, y) = 0$ *has infinitely many rational solutions with a bounded denominator* if there exists a positive integer Δ such that $F(x, y) = 0$ has infinitely many solutions $(x, y) \in \mathbb{Q} \times \mathbb{Q}$ with $\Delta x, \Delta y \in \mathbb{Z}$.

Before stating our criterion, we consider several examples of equation (1.1) having infinitely many solutions. In what follows a and b are non-zero rational numbers, m and n are positive integers, and $p(x) \in \mathbb{Q}[x]$ is a non-zero polynomial (which may be constant).

Example 1.1. The equation $x^m = ay^n$ with coprime m and n has infinitely many solutions with the bounded denominator. Indeed, if the positive integers μ and ν are such that $\mu m - \nu n = 1$ then $x = a^\mu t^n$ and $y = a^\nu t^m$ is a solution for every $t \in \mathbb{Z}$. Moreover, every solution (x, y) of $x^m = ay^n$ produces a solution $(xp(y), y)$ of the more general equation

$$x^m = ay^n p(y)^m, \tag{1.2}$$

and so this latter equation has infinitely many solutions with a bounded denominator as well.

Example 1.2. The *Pell equation* $x^2 = ay^2 + b$ is known to have infinitely many solutions with a bounded denominator for (many) suitable choices of a and b. Moreover, each solution (x, y) of this equation produces a solution $(xp(y), y)$ of the more general equation

$$x^2 = (ay^2 + b)p(y)^2, \tag{1.3}$$

so the latter equation has infinitely many solutions with a bounded denominator as well.

Example 1.3. Recall that the *Chebyshev polynomial* $T_n(x)$ is defined by the functional equation $T_n(\cos z) = \cos nz$, or, alternatively, by $T_n((z + z^{-1})/2) = (z^n + z^{-n})/2$. It follows from the definition that $T_n \circ T_m = T_m \circ T_n = T_{mn}$. Hence the equation $T_m(x) = T_n(y)$ has infinitely solutions with the bounded denominator given by $x = T_n(t)$ and $y = T_m(t)$, where $t \in \mathbb{Z}$.

A useful modification of Chebyshev polynomials are *Dickson polynomials*, which depend on an additional parameter a. The Dickson polynomial $D_n(x, a)$ is defined by $D_n(z + a/z, a) = z^n + (a/z)^n$, and we have $D_n(x, 1) = 2T_n(x/2)$. It follows from the definition that

$$D_m\left(D_n(x, a), a^n\right) = D_n\left(D_m(x, a), a^m\right) = D_{mn}(x, a).$$

Hence the equation

$$D_m(x, a^n) = D_n(y, a^m) \tag{1.4}$$

has infinitely many solutions with the bounded denominator given by $x = D_n(t, a)$ and $y = D_m(t, a)$.

Example 1.4. Our next example is the equation

$$a^{-m/2} D_m(x, a) = -b^{-n/2} D_n(y, b), \tag{1.5}$$

where $\gcd(m, n) = 2$. Define the polynomial $E_n(x, a)$ by the relation $E_n(z + a/z, a) = (z^n - (a/z)^n)/(z - a/z)$. One can verify that every solution (u, v) of the Pell equation $a^{m/2} u^2 + bv^2 = 4ab$ gives rise to a solution[2] $x = a^{(2-n)/4} D_{n/2}(v, a)$ and $y = u E_{m/2}(v, a)$ of (1.5).

Example 1.5. Our final example is the equation

$$(ax^2 - 1)^3 = 3y^4 - 4y^3. \tag{1.6}$$

A quick inspection shows that a solution of the Pell equation $3au^2 = v^2 + 2$ gives rise to a solution $x = u(v + 2)$ and $y = ((v + 1)^3 + 4)/3$ of (1.6).

We make also two simple observations.

Observation 1.6. If the equation $f(x) = g(y)$ has infinitely many solutions with a bounded denominator, and $\lambda(x), \mu(x) \in \mathbb{Q}[x]$ are non-constant linear polynomials, then the equation $f \circ \lambda(x) = g \circ \lambda(y)$ has infinitely many solutions with a bounded denominator as well.

Observation 1.7. If the equation $f(x) = g(y)$ has infinitely many solutions with a bounded denominator, and $\varphi(x) \in \mathbb{Q}[x]$ is an arbitrary polynomial, then the equation $\varphi \circ f(x) = \varphi \circ g(y)$ has infinitely many solutions with a bounded denominator as well.

Our main result asserts that all possible equations $f(x) = g(y)$ having infinitely many solutions with a bounded denominator can be obtained from the examples and observations above.

Theorem 1.8. *Let $f(x), g(x) \in \mathbb{Q}[x]$ be non-constant polynomials such the equation $f(x) = g(y)$ has infinitely many rational solutions with a bounded denominator. Then $f = \varphi \circ f_1 \circ \lambda$ and $g = \varphi \circ g_1 \circ \mu$, where $\lambda(x), \mu(x) \in \mathbb{Q}[x]$ are non-constant linear polynomials, $\varphi(x) \in \mathbb{Q}[x]$,*

---

[2] Without loss of generality we may assume that $n/2$ is odd.

and the pair $(f_1(x), g_1(x))$ *is (up to switching f_1 and g_1) of one the five types*

$$\left(x^m, ax^r p(x)^m\right), \quad \left(x^2, (ax^2 + b)p(x)^2\right), \quad \left(D_m(x, a^n), D_n(x, a^m)\right),$$
$$\left(a^{-m/2} D_m(x, a), -b^{-n/2} D_n(x, b)\right), \quad \left((ax^2 - 1)^3, 3x^4 - 4x^3\right).$$

occurring in the examples above.

Actually, we obtain a more general result, on solutions of (1.1) in S-integers of an arbitrary number field. In this case one more example occurs, besides the five given above.

Finiteness conditions for the equation (1.1) were studied by many authors. For the equation $f(x) = y^n$ a finiteness theorem was established by Siegel [31] and Le Veque [23]. Evertse and Silverman [15] gave a sharp estimate for the number of solutions. Starting from Baker [3], many authors (notably, Sprindzhuk, Brindza and Bugeaud) developped methods for the effective analysis of the equation $f(x) = y^n$; see [10,33] for the references. An algorithm for the numerical solution of this equation was suggested in [7].

Davenport, Lewis and Schinzel [12] obtained a finiteness condition for the general equation (1.1). However, it was too restrictive for many applications.

Fried investigated the problem from various points of view in a remarkable series of paper [16–18]. In particular, he gave in [18, Corollary after Theorem 3] a new finiteness condition, much more general then that of [12], but still not completely explicit.

Schinzel [29, Theorem 8] obtained a completely explicit finiteness criterion for the equation (1.1) under the assumption

$$\gcd(\deg f, \deg g) = 1.$$

(Notice that the latter assumption is quite restrictive; in particular, it implies that the polynomial $f(x) - g(y)$ is irreducible.) Schinzel showed that in this case $f = \varphi \circ f_1 \circ \lambda$ and $g = \varphi \circ g_1 \circ \mu$ with linear polynomials φ, λ and μ and with f_1, g_1 as in equations (1.2) or (1.4). This also follows from our theorem.

The proof of Theorem 1.8 follows the main lines of the arguments of Fried [18] and Schinzel [29]. However, several substantially new ideas were required to drop Schinzel's assumption $(\deg f, \deg g) = 1$, retaining the explicit character of his result.

Call an absolutely irreducible polynomial $F(x, y) \in \mathbb{Q}[x, y]$ *exceptional* if the corresponding plane curve is of genus 0 and has at most 2 points at infinity. To deduce Theorem 1.8 from Siegel's theorem, one has to determine when the polynomial $f(x) - g(y)$ has an exceptional factor. This reduces our original problem, which is of arithmetical nature, to a purely algebraic problem; in fact, the main part of the argument works over an arbitrary field of characteristic 0.

Our argument consists of two parts. In the first part we assume that the polynomial $f(x) - g(y)$ is irreducible, in which case it is exceptional itself. Since the number of points at infinity is equal to $\gcd(\deg f, \deg g)$, we have the following problem: classify pairs (f, g) such that $\gcd(\deg f, \deg g) = 1$ or 2 and such that the curve $f(x) = g(y)$ is of genus 0. The case $\gcd = 1$ is the classical "Second theorem of Ritt" [28], in the form presented by Schinzel in [29, 30]. In the case $\gcd = 2$ the argument is similar, but technically much more complicated. The proof relies on quite tricky computations involving the Riemann-Hurwitz genus formula.

In the second part we drop the irreducibility assumption and consider the general case. Due to a clever observation of Fried, the problem, basically, reduces to the case when $\deg f = \deg g$. It is easy to see that if $q(x, y)$ divides $f(x) - g(y)$ with $\deg f = \deg g$, then the curve $q(x, y) = 0$ has exactly $\deg q$ points at infinity. In particular, if this polynomial $q(x, y)$ is exceptional as defined above, then $\deg q \leq 2$. Thus, it remains to classify pairs (f, g) such that $f(x) - g(y)$ has a factor of degree at most 2. This problem is is completely solved in [5] (again relying upon the ideas of Fried).

After all this work is done, Theorem 1.8 becomes an (almost) immediate consequence of Siegel's theorem.

Extensions, generalizations and applications of Theorem 1.8 can be found in [1, 2, 6, 8, 11, 13, 14, 19–22, 24–27, 34–38].

References

[1] R. M. AVANZI and U. M. ZANNIER, *Genus one curves defined by separated variable polynomials and a polynomial Pell equation*, Acta Arith. **99** (2001), 227–256.

[2] R. M. AVANZI and U. M. ZANNIER, *The equation $f(X) = f(Y)$ in rational functions $X = X(t)$, $Y = Y(t)$*, Compositio Math. **139** (2003), 263–295.

[3] A. BAKER, *Bounds for solutions of superelliptic equations*, Proc. Cambridge Phil. Soc. **65** (1969), 439–444.

[4] YU. BILU, *Integral points and Galois covers*, Math. Contemp. **14** (1998), 1–11.

[5] YU. BILU, *Quadratic factors of* $f(x) - g(y)$, Acta Arith. **90** (1999), 341–355.

[6] YU. F. BILU, B. BRINDZA, P. KIRSCHENHOFER, A. PINTÈR and R. F. TICHY, *Diophantine equations and Bernoulli polynomials* (with an appendix by A. Schinzel), Compositio Math. **131** (2002), 173–188.

[7] YU. BILU and G. HANROT, *Solving superelliptic Diophantine equations by Baker's method*, Compositio Math. **112** (1998), 273–312.

[8] YU. F. BILU, TH. STOLL and R. F. TICHY, *Octahedrons with equally many lattice points*, Period. Math. Hungar. **40** (2000), 229–238.

[9] YU. F. BILU and R. F. TICHY, *The Diophantine equation* $f(x) = g(y)$, Acta Arith. **95** (2000), 261–288.

[10] Y. BUGEAUD, *Bounds for the solutions of superelliptic equations*, Compositio Math. **107** (1997), 187–219.

[11] Y. BUGEAUD and F. LUCA, *On Pillai's Diophantine equation*, New York J. Math. **12** (2006), 193–217 (electronic).

[12] H. DAVENPORT, D. J. LEWIS and A. SCHINZEL, *Equations of the form* $f(x) = g(y)$, Quat. J. Math. Oxford **12** (1961), 304–312.

[13] A. DUJELLA and I. GUSIĆ, *Indecomposability of polynomials and related Diophantine equations*, Q. J. Math. **57** (2006), 193–201.

[14] A. DUJELLA and R. F. TICHY, *Diophantine equations for second-order recursive sequences of polynomials*, Q. J. Math. **52** (2001), 161–169.

[15] J.-H. EVERTSE and J. H. SILVERMAN, *Uniform bounds for the number of solutions to* $Y^n = f(X)$, Math. Proc. Cambridge Philos. Soc. **100** (1986), 237–248.

[16] M. FRIED, *The field of definition of function fields and a problem in the reducibility of polynomials in two variables*, Ill. J. Math. **17** (1973), 128–146.

[17] M. FRIED, *Arithmetical properties of function fields. II. The generalized Schur problem*, Acta Arith. **25** (1973/74), 225–258.

[18] M. FRIED, *On a theorem of Ritt and related Diophantine problems*, J. reine angew. Math. **264** (1974), 40–55.

[19] C. FUCHS, A. PETH and R. F. TICHY, *On the Diophantine equation* $G_n(x) = G_m(P(x))$, Monatsh. Math. **137** (2002), 173–196.

[20] M. KULKARNI and B. SURY, *On the Diophantine equation* $x(x + 1)(x + 2) \ldots (x + (m - 1)) = g(y)$, Indag. Math. (N.S.) **14** (2003), 35–44.

[21] M. KULKARNI and B. SURY, *A class of Diophantine equations involving Bernoulli polynomials*, Indag. Math. (N.S.) **16** (2005), 51–65.

[22] M. KULKARNI and B. SURY, *Diophantine equations with Bernoulli polynomials*, Acta Arith. **116** (2005), 25–34.

[23] LE VEQUE, *On the equation $y^m = f(x)$*, Acta Arith. **9** (1964), 209–219.

[24] F. LUCA, *Arithmetic properties of members of a binary recurrent sequence*, Acta Arith. **109** (2003), 81–107.

[25] F. LUCA, *On the Diophantine equation $p^{x_1} - p^{x_2} = q^{y_1} - q^{y_2}$*, Indag. Math. (N.S.) **14** (2003), 207–222.

[26] C. RAKACZKI, *On the Diophantine equation $x(x-1)\cdots(x-(m-1)) = \lambda y(y-1)\cdots(y-(n-1))+l$*, Acta Arith. **110** (2003), 339–360.

[27] C. RAKACZKI, *On the Diophantine equation $S_m(x) = g(y)$*, Publ. Math. Debrecen **65** (2004), 439–460.

[28] J. F. RITT, *Prime and composite polynomials*, Trans. Amer. Math. Soc. **23** (1922), 51–66.

[29] A. SCHINZEL, "Selected Topics on Polynomials", The University of Michigan Press, Ann Arbor, 1982.

[30] A SCHINZEL, "Polynomials with Special Regard to Reducibility" (with an appendix by U. Zannier), Encyclopedia of Mathematics and its Applications **77**, Cambridge University Press, Cambridge, 2000.

[31] C. L. SIEGEL, *The integer solutions of the equation $y^2 = ax^n + bx^{n-1} + \cdots + k$*, J. London Math. Soc., **1** (1926), 66–68; Ges. Abh. Band 1, 207–208.

[32] C. L. SIEGEL, *Über einige Anwendungen Diophantischer Approximationen*, Abh. Preuss Akad. Wiss. Phys.-Math. Kl. **1** (1929); Ges. Abh. Band 1, 209–266.

[33] V. G. SPRINDŽUK, "Classical Diophantine Equations in Two Unknowns" (Russian), Nauka, Moscow, 1982; English trans.: "Lecture Notes in Math.", Vol. 1559, Springer, 1994.

[34] TH. STOLL, *Diophantine equations for orthogonal polynomials*, Riv. Mat. Univ. Parma (7) **3*** (2004), 311–321.

[35] TH. STOLL and R. F. TICHY, *Diophantine equations for classical continuous orthogonal polynomials*, Indag. Math. (N.S.) **14** (2003), 263–274.

[36] TH. STOLL and R. F. TICHY, *The Diophantine equation $\alpha\binom{x}{m} + \beta\binom{y}{n} = \gamma$*, Publ. Math. Debrecen **64** (2004), 155–165.

[37] TH. STOLL and R. F. TICHY, *Diophantine equations involving general Meixner and Krawtchouk polynomials*, Quaest. Math. **28** (2005), 105–115.

[38] SZ. TENGELY, *On the Diophantine equation $F(x) = G(y)$*, Acta Arith. **110** (2003), 185–200.

Hopf algebras with trace and Clebsch-Gordan coefficients

Corrado De Concini

Abstract. In this lecture I shall report on some joint work with Procesi, Reshetikhin and Rosso [1].

1. Recollections and the problem

Let G be a simply connected semisimple algebraic group over $\mathbb{C}$. $\mathfrak{g} = \mathrm{Lie}\, G$.

$T \subset G$ a maximal torus, $\mathfrak{t} = \mathrm{Lie}\, T$, R the root system, R_+ positive roots, $\Delta = \{\alpha_1, \ldots, \alpha_n\}$ the simple roots, $A = (a_{i,j})$, the Cartan matrix, $D = \mathrm{diag}(d_1, \ldots, d_n)$ the diagonal matrix with DA symmetric. $B \supset T \subset B^-$ the corresponding Borel subgroup and its opposite.

Fix ℓ odd (and prime with 3 if there are G_2 components). Let ε be a primitive ℓ root of 1.

The quantized enveloping algebra is the $\mathbb{C}$-algebra $U_\varepsilon(\mathfrak{g})$ generated by elements $\{E_1, \ldots E_n\}$, $\{F_1, \ldots F_n\}$, K_λ, $\lambda \in X^*(T)$ the character group of T, with relations:

$$K_\lambda K_\mu = K_{\lambda+\mu}$$
$$K_\lambda E_i K_\lambda^{-1} = \varepsilon^{\langle \check{\alpha}_i, \lambda \rangle} E_i$$
$$K_\lambda F_i K_\lambda^{-1} = \varepsilon^{-\langle \check{\alpha}_i, \lambda \rangle} F_i$$

$$E_i F_j = F_j E_i \quad i \neq j, \quad [E_i, F_i] = \frac{K_{\alpha_i} - K_{\alpha_i}^{-1}}{\varepsilon_i - \varepsilon_i^{-1}} \quad \varepsilon_i = \varepsilon^{d_i}$$

$$\sum \begin{bmatrix} 1 - a_{ij} \\ k \end{bmatrix}_{\varepsilon_i} E_i^{1-k-a_{ij}} E_j E_i^k = 0, \quad i \neq j$$

$$\sum \begin{bmatrix} 1 - a_{ij} \\ k \end{bmatrix}_{\varepsilon_i} F_i^{1-k-a_{ij}} F_j F_i^k = 0, \quad i \neq j$$

$$\begin{bmatrix} m \\ h \end{bmatrix}_\varepsilon = \frac{[m]_\varepsilon!}{[m-h]_\varepsilon![h]_\varepsilon!}, \quad [h]_\varepsilon! = [h]_\varepsilon \ldots [2]_\varepsilon [1]_\varepsilon \text{ and } [h]_\varepsilon = \frac{\varepsilon^h - \varepsilon^{-h}}{\varepsilon - \varepsilon^{-1}}. \; \Delta \text{ acts}$$

on generators by

$$\Delta K_\mu = K_\mu \otimes K_\mu, \quad \Delta E_i = E_i \otimes 1 + K_{\alpha_i} \otimes E_i,$$
$$\Delta F_i = F_i \otimes K_{\alpha_i}^{-1} + 1 \otimes F_i.$$

The main peculiarity of being at roots of unity is that $U_\varepsilon(\mathfrak{g})$ has a very big center Z.

Indeed, $U_\varepsilon(\mathfrak{g})$ is a finite Z-module. Z is the coordinate ring of an algebraic variety X of dimension equal to dim $\mathfrak{g}$. So:

1. Every irreducible $U_\varepsilon(\mathfrak{g})$-module is finite dimensional.
2. If $\widehat{U_\varepsilon(\mathfrak{g})}$ denotes the set of irreducible $U_\varepsilon(\mathfrak{g})$-modules, taking central characters we get a surjective map

$$\gamma : \widehat{U_\varepsilon(\mathfrak{g})} \to X.$$

Problem 1.1. Given two irreducible $U_\varepsilon(\mathfrak{g})$-modules, V, W, describe the composition factors of $V \otimes W$ and their multiplicities.

Consider the sub Hopf algebra $U_\varepsilon(\mathfrak{b}) \subset U_\varepsilon(\mathfrak{g})$ generated by the K_λ and $E_1, \ldots E_n$.

Problem 1.2. Given an irreducible $U_\varepsilon(\mathfrak{g})$-module, V "decompose" its restriction to $U_\varepsilon(\mathfrak{b})$.

Notice that as Problem 1.2, also Problem 1.1 is a problem of branching. In fact it consists of decomposing the restriction of the irreducible $U_\varepsilon(\mathfrak{g}) \otimes U_\varepsilon(\mathfrak{g})$-module $V \otimes W$ to the subalgebra $\Delta(U_\varepsilon(\mathfrak{g}))$.

To explain our results recall that Z contains a sub Hopf algebra Z_0 such that $U_\varepsilon(\mathfrak{g})$ is a free Z_0 module of rank $\ell^{\dim \mathfrak{g}}$. Z_0 is the coordinate ring of an algebraic group H (the Poisson dual of G) which is the kernel of the homomorphism $p : B \times B^- \to T$ defined by $p((b, b')) = p_+(b)p_-(b')^{-1}$ where $p_\pm$ are the quotients modulo the unipotent radicals. The inclusion $Z_0 \subset Z$ gives a map $\pi : X \to H$.

Theorem 1.3. *There is a non empty Zariski open set $V \subset H \times H$ such that if V and W are two irreducible representations of $U_\varepsilon(\mathfrak{g})$, such that if $(h_1, h_2) = (\pi\gamma(V), \pi\gamma(W)) \in V$, as a $U_\varepsilon(\mathfrak{g})$ module,*

$$V \otimes W = \oplus_{U \in (\pi\gamma)^{-1}(h_1 h_2)} U^m$$

with $m = \ell^{|R_+|-\mathrm{rk}\mathfrak{g}}$.

As for Problem 1.2, $Z_0^+ = Z_0 \cap U_\varepsilon(\mathfrak{b})$ is a sub Hopf algebra of $U_\varepsilon(\mathfrak{b})$. Z_0^+ is the coordinate ring of B^- and the inclusion $Z_0^+ \subset Z_0$ induces the homomorphism $\mu : H \to B^-$ given by the projection on the second factor. Again, taking central character we get a map

$$\gamma' : \widehat{U_\varepsilon(\mathfrak{b})} \to B^-.$$

Theorem 1.4. *There is a non empty Zariski open set $V \subset H$ such that if V is an irreducible representations of $U_\varepsilon(\mathfrak{g})$, such that if $h = \pi\gamma(V) \in V$, as a $U_\varepsilon(\mathfrak{b})$ module,*

$$V = \oplus_{U \in (\gamma')^{-1}(\mu(h))} U^m$$

with $m = \ell^{(|R_+|-s)/2}$ where s is the number of orbits of $-w_0$ on the set of simple roots.

The main content of these theorems is that in both cases, at least generically, the multiplicities are uniformly distributed among the various irreducible components.

2. Cayley-Hamilton algebras

The results we have just stated follow from a careful study of the centers of the algebras involved, by applying the theory of Cayley-Hamilton algebras which we now briefly recall.

A trace on a algebra R over a field k (here $= \mathbb{C}$) is a 1-ary operation $t : R \to R$ such that

1. t is $k-$linear. (This implies that $t(R)$ is an $k-$subspace),
2. $t(a)b = bt(a)$, $\quad \forall a, b \in R$. (This implies that $t(R)$ is in the center of R),
3. $t(ab) = t(ba)$, $\quad \forall a, b \in R$. (This implies that t is 0 on the space of commutators $[R, R]$),
4. $t(t(a)b) = t(a)t(b)$, $\quad \forall a, b \in R$. (This implies that $t(R)$ is an subalgebra, called the trace algebra, and that t is also $t(R)-$linear).

The most famous example is that of the algebra M_n of $n \times n$ matrices with the usual trace tr.

Algebras with trace form a category. Given (R_1, t_1), (R_2, t_2) a morphism ϕ from (R_1, t_1) to (R_2, t_2) is an algebra homomorphism such that $\phi t_1(a) = t_2\phi(a)$ for each $x \in R_1$.

A morphism $\phi : (R, t) \to (M_n, \text{tr})$ is called a representation.

(M_n, tr) have the following extra properties:

1. $\text{tr}(1) = n$,
2. given $A \in M_n$, set $p_a^{(n)}(x) = \det(xI - A)$, the characteristic polynomial, then we get the Cayley-Hamilton identity

$$p_a^{(n)}(a) = 0.$$

Remark that $p_a^{(n)}(x)$ has coefficients which are universal polynomials in $\text{tr}(a), \ldots, \text{tr}(a^n)$. For example for $n = 2$,

$$p_a^{(2)}(x) = x^2 - \text{tr}(a)x + \frac{1}{2}(\text{tr}(a)^2 - \text{tr}(a^2)).$$

Thus, given (R, t) we can consider $p_a^{(n)}(x)$ for every $a \in R$ as a polynomial with coefficients in the trace algebra and define

Definition 2.1. An algebra with trace (A, t) is a n-Cayley Hamilton algebra (or briefly an n-CH algebra) if

1. $t(1) = n$
2. $p_a^{(n)}(a) = 0$ for all $a \in R$.

Remark that if $\rho : (R, t) \to (M_r, \text{tr})$ is a representation of a n-CH algebra, then $r = n$ since

$$r = \text{tr}(1) = \text{tr}(\rho(1)) = \rho(t(1)) = \rho(n) = n.$$

If (R, t) n-CH algebra and h is a positive integer, (R, ht) is a hn-CH algebra and if V is a (n dimensional) (R, t)-module, $V^{\oplus h}$ is a (R, ht)-module.

The category of C-H algebras clearly contains free objects. The free n-CH algebra $F_n\langle x_1, \ldots, x_m \rangle$ on m generators is the algebra generated freely by the x_i's and by traces of monomials modulo the ideal generated by evaluating the nth Cayley Hamilton identity.

3. Representations of Cayley-Hamilton algebras

The following results describes the free n-CH algebra $F_n\langle x_1, \ldots, x_m \rangle$. Consider the algebra $C_n(\xi_1, \ldots, \xi_m)$ of $GL(n)$-equivariant polynomial maps of matrices $M_n(k)^m \to M_n(k)$, $\xi_i(A_1, \ldots, A_m) = A_i$, with trace $t(\psi) = \text{tr} \cdot \psi$. We have a universal map $F_n\langle x_1, \ldots, x_m \rangle \to C_n(\xi_1, \ldots, \xi_m)$.

Theorem 3.1 (Procesi). 1. *The universal map*

$$F_n\langle x_1, \ldots, x_m \rangle \xrightarrow{\ i\ } C_n(\xi_1, \ldots, \xi_m)$$

is an isomorphism.
2. *The trace algebra $T_n(\xi_1, \ldots, \xi_m)$ of the algebra $C_n(\xi_1, \ldots, \xi_m)$ is the algebra of invariants of m-tuples of matrices and, if $m > 1$, it is the center of $C_n(\xi_1, \ldots, \xi_m)$).*
3. *$C_n(\xi_1, \ldots, \xi_m)$ is a finite $T_n(\xi_1, \ldots, \xi_m)$ module.*

The above theorem tells us in particular that if (R, t) is a n-CH algebra which is finitely generated also its trace algebra A is finitely generated and R is a finite A-module.

It follows that every irreducible representation of R (as an algebra) is finite dimensional.

A is a ring of functions on an affine algebraic variety X, so if $\widehat{R}$ denotes the set of irreducible representations of R, we get, by taking central characters, a surjective map

$$\gamma : \widehat{R} \to X.$$

The points of X itself are a "moduli space" for a class of representations. Indeed

Theorem 3.2. *Given a point $x \in X$, there is a unique semisimple, trace compatible, representation W_x of the form*

$$W_x = \oplus_{V \in \gamma^{-1}(x)} V^{\oplus m_V},$$

and each semisimple, trace compatible, representation of R is of this form.

In fact there is a Zariski open set (possibly empty), called the unramified locus, in X such that if $m_x \subset A$ is the maximal ideal in A, $R/m_x R$ is semisimple and so W_x is the unique representation "lying over x" which is compatible with the trace.

If A is the center of R and R satisfies suitable conditions, for example it is a domain, there is a non empty Zariski open set for which W_x is irreducible.

In general however, the determination of $\gamma^{-1}(x)$ and of the multiplicities m_V is a very hard problem.

If we consider (R, ht), with h a positive integer, then the semisimple, trace compatible, representation corresponding to $x \in X$ with respect to the new trace is just $W_x^{\oplus h}$.

4. Reduced trace

In our example of quantized enveloping algebra we have no trace, so if we want to use the results about the CH algebras we need to find ways to canonically introduce a trace.

Here is a way of doing this under some assumptions.

A prime algebra R is an algebra in which the product of two non-zero ideals is non-zero. Let R be a prime algebra, assume that A is a subalgebra of the center and R is an A-module of finite type. Then:

1. A is an integral domain.
2. R is a torsion free module.

So if F is the field of fractions of A, then $R \subset R \otimes_A F$ and $S := R \otimes_A F$ is a finite dimensional simple algebra isomorphic to $M_k(D)$ with D a finite dimensional division ring.

Denote by Z the center of S and of D. Then $\dim_Z D = h^2$.

If $\overline{Z}$ is an algebraic closure of Z, $M_k(D) \otimes_Z \overline{Z} = M_{hk}(\overline{Z})$. Setting $p := [Z : F] = \dim_F Z$, we also have

$$S \otimes_F \overline{Z} = M_k(D) \otimes_F \overline{Z} = M_{hk}(\overline{Z})^{\oplus p}.$$

The number $n := hkp$ is called *degree* of S over F. Consider the F-linear operator $a^L : S \to S$, $a^L(b) := ab$. We define the reduced trace

$$t_{S/F}(a) = \frac{1}{hk} \operatorname{tr}(a^L).$$

This theorem tells us that, provided A is integrally closed, we have constructed a nice trace:

Theorem 4.1. *If $S = R \otimes_A F$ as before and A is integrally closed we have that the reduced trace $t_{S/F}$ maps R into A, so we will denote by $t_{R/A}$ the induced trace.*

The algebras R and S with their reduced trace are n-Cayley Hamilton algebras.

Even more important for us is the fact that, in the situation above, the nice trace is essentially unique. Indeed,

Theorem 4.2. *Under the same circumstances of Theorem 4.1, if $\tau : R \to A$ is any trace for which R is an m-Cayley Hamilton algebra then there is a positive integer r for which:*

$$m = rn, \qquad \tau = r\, t_{R/A}$$

Notice that in particular, this implies that if x is a point in X and W_x is the unique semisimple representation above x, compatible with $t_{R/A}$, then $W_x^{\oplus r}$ is the unique semisimple representation above x compatible with τ.

5. Compatibility

Since our final goal is to analyze the restriction of certain representation of an algebra R to a subalgebra, what I am going to explain now is how the reduced trace of R relative to a subring A restricts to a subalgebra.

Assume that we have two prime algebras $R_1 \subset R_2$ with two central subrings $A_1 \subset R_1$, $A_2 \subset R_2$, $A_1 \subset A_2$ such that R_1 is a finite A_1 module, R_2 is a finite A_2 module and A_1, A_2 are integrally closed. We have the reduced traces t_{R_1/A_1} and t_{R_2/A_2} for which both algebras are Cayley-Hamilton. We need a criterion ensuring that the restriction of t_{R_2/A_2} to R_1 is a multiple of t_{R_1/A_1} (in this case we say that they are compatible). The criterion is the following,

Theorem 5.1. *Denote by Z_1 the center of R_1. If the algebra $Z_1 \otimes_{A_1} A_2$ is a domain, there is a positive integer r such that*

$$r\, t_{R_1/A_1} = t_{R_2/A_2} \quad on \ R_1.$$

6. Back to quantum algebras

In this final section I am going to explain how the theory of Cayley Hamilton algebras can be applied to the quantum group situation described in Section 1.

As a first step we need to show that quantized enveloping algebras have reduced traces. This is not hard. One knows that both $U_\varepsilon(\mathfrak{g})$ and $U_\varepsilon(\mathfrak{b})$ are domains and hence prime and that their centers Z, Z' are integrally closed. Z_0 and Z_0^+ are clearly integrally closed, so we have reduced traces

$$t_{U_\varepsilon(\mathfrak{g})/Z_0} \quad t_{U_\varepsilon(\mathfrak{g})/Z'} \quad t_{U_\varepsilon(\mathfrak{g})^{\otimes 2}/Z^{\otimes 2}} \quad t_{U_\varepsilon(\mathfrak{b})/Z_0^+}.$$

Next we have to compare them.

We start with Problem 1.1 and the inclusion $\Delta(U_\varepsilon(\mathfrak{g})) \subset U_\varepsilon(\mathfrak{g}) \otimes U_\varepsilon(\mathfrak{g})$. Unfortunately we have that $\Delta(Z)$ is not contained in $Z \otimes Z$. On the other hand, the fact that Z_0 is a sub Hopf algebra means that $\Delta(Z_0) \subset Z_0 \otimes Z_0 \subset Z \otimes Z$. So, we may try to compare $t_{U_\varepsilon(\mathfrak{g})^{\otimes 2}/Z^{\otimes 2}}$ with $t_{U_\varepsilon(\mathfrak{g})/Z_0}$.

In order to use Theorem 5.1, we need to show that the variety $\tilde{X}$ defined as the fiber product

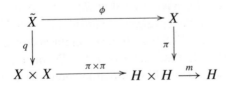

with $m : H \times H \to H$ the group multiplication, is irreducible.

Let us recall that X is obtained as the fiber product

$$
\begin{array}{ccc}
X & \xrightarrow{\;\;p\;\;} & T/W \\
\downarrow{\scriptstyle\pi} & & \downarrow{\scriptstyle\ell} \\
H & \xrightarrow{\;\;\rho\;\;} G \xrightarrow{\;\sigma\;} & G//G \cong T/W.
\end{array}
$$

Where for $(h, k) \in H$, $\rho((h, k)) = hk^{-1}$, $G//G$ denotes the categorical quotient modulo the adjoint action defined using invariant polynomial functions, σ is the quotient map, W is the Weyl group and the map ℓ is the map induced by the homomorphism $t \to t^\ell$ of the torus T on W-invariants.

Putting the two diagrams together we obtain $\tilde{X}$ as the fiber product

$$
\begin{array}{ccc}
\tilde{X} & \xrightarrow{\;\;p\phi\;\;} & T/W \\
\downarrow{\scriptstyle q} & & \downarrow{\scriptstyle\ell} \\
X \times X & \xrightarrow{\;\sigma \cdot \rho \cdot m \cdot \pi \times \pi\;} & T/W.
\end{array}
$$

From this one proves:

STEP 1. $\tilde{X}$ is a complete intersection over $X \times X$ which is Cohen-Macaulay so it is Cohen-Macaulay.

STEP 2. $\tilde{X}$ is non singular in codimension 1 so in particular it is reduced and normal.

STEP 3. $\tilde{X}$ is connected. This is the hardest step and one needs to use a variation of Steinberg section for the quotient $\sigma \colon G \to G//G$.

Using the irreducibility of $\tilde{X}$, we deduce that the restriction to $\Delta(U_\varepsilon(\mathfrak{g}))$ of $t_{U_\varepsilon(\mathfrak{g})^{\otimes 2}/Z^{\otimes 2}}$ is $m t_{U_\varepsilon(\mathfrak{g})/Z_0}$ with $m = \ell^{|R_+|-\mathrm{rk}\mathfrak{g}}$ and Theorem 1.3 follows.

The open set $\mathcal{V} \subset H \times H$ consists of the set of pairs $((h_1, k_1), (h_2, k_2))$ with $h_1 k_1^{-1}, h_2 k_2^{-1}, h_1 h_2 k_2^{-1} k_1^{-1}$ all regular semisimple.

The solution to Problem 1.2 is quite similar with the extra difficulty that one has to compute the center of $U_\varepsilon(\mathfrak{b})$ first since this has not been determined before. We have seen that Z_0^+ is the coordinate ring of the group B^-. For a dominant weight λ, consider the irreducible G-module V_λ of highest weight λ. Take a highest weight vector $v \in V_\lambda$, a highest weight vector ϕ and a lowest vector ψ in V_λ^*. Define the function $f_\lambda(b) = \phi(bv)\psi(bv)$ for $b \in B^-$. These functions generate a subring $\mathcal{Z}$ in Z_0^+, isomorphic to the polynomial ring $\mathbb{C}[f_{\omega_1}, \ldots, f_{\omega_r}]$, ω_i the fundamental weights.

Embed $\mathcal{Z}$ into the polynomial ring

$$\mathcal{Z}' = \mathbb{C}[f_{\omega_1}^{1/\ell}, \ldots, f_{\omega_r}^{1/\ell}].$$

$\mathcal{Z}$ is the ring of invariants in $\mathcal{Z}'$ under an obvious action of the group $\mathbb{Z}/\ell\mathbb{Z}$ which we can consider as the group T_ℓ of ℓ-torsion points in T. Set $\Gamma = \{t \in T_\ell \,|\, t^{w_0} = t^{-1}\}$, and $Z_1^+ = (\mathcal{Z}')^\Gamma$.

Theorem 6.1. *The center of the algebra $U_\varepsilon(\mathfrak{b})$ is isomorphic to $Z^+ = Z_0^+ \otimes_{\mathcal{Z}} Z_1^+$.*

Once we know the center of $U_\varepsilon(\mathfrak{b})$, in order to solve Problem 1.2 we need to compare $t_{U_\varepsilon(\mathfrak{g})/Z}$ and $t_{U_\varepsilon(\mathfrak{b})/Z_0^+}$. In order to use Theorem 5.1 we need to show that the fiber product $\tilde{Y}$

$$
\begin{array}{ccc}
\tilde{Y} & \longrightarrow & Y \\
{\scriptstyle q}\downarrow & & \downarrow{\scriptstyle \ell} \\
X & \xrightarrow{\ \pi\ } & H \xrightarrow{\ p_-\ } B^-
\end{array}
$$

with $Y = \operatorname{Spec} Z^+$, is a irreducible variety. This is done as in the previous case. After these facts have been established, the proof of Theorem 1.4 follows rather easily.

7. Final remarks and problems

The technique explained can be used in other cases. One example is the so called quantized function algebra $\mathbb{F}_\varepsilon[G]$. In this case the role played by H is played by G and we have a projection $\pi : \widehat{F_\varepsilon[G]} \to G$.

Theorem 7.1. *There is a non empty Zariski open set $V \subset G \times G$ such that if V and W are two irreducible representations of $\mathbb{F}_\varepsilon[G]$, such that if $(g_1, g_2) = (\pi(V), \pi(W)) \in V$, as a $\mathbb{F}_\varepsilon[G]$ module,*

$$V \otimes W = \oplus_{U \in (\pi)^{-1}(g_1 g_2)} U^m$$

with $m = \ell^{|R_+|-\operatorname{rk}\mathfrak{g}}$.

References

[1] C. DE CONCINI, C PROCESI, N. RESHETIKHIN and M. ROSSO, *Hopf algebras with trace and representations*, Invent. Math. **161** (2005), 1–44.

The integral Cauchy formula on symmetric Stein manifolds

Simon Gindikin

Abstract. We discuss the harmonic analysis on complex semisimple Lie groups and, more general, on symmetric Stein manifolds from point of view of multidimensional complex analysis.

Harmonic analysis on complex semisimple Lie groups was constructed by E. Cartan and H. Weyl using 2 approaches: algebraic and analytic (transcendental). They permanently compared these methods. The algebraic way of E. Cartan uses Lie algebras. The transcendental method of H. Weyl is non direct: he had no appropriate analytic tools on complex groups G. Instead he works with their compact forms U and develops on them a version of the Peter-Weyl construction using the invariant integration. The crucial point is the famous unitary trick of H. Weyl. It turns out that irreducible representations of a group G and its maximal compact subgroup U coincide. The analytic nature of this a phenomenon was not clarified for a log time. A possible explanation is that U is the Riemannian manifold and invariant differential operators on U are elliptic. Irreducible representations are realized in joint eigenspaces of them. It turns out that the eigenfunctions not only analytic on U but extend holomorphically on the whole complex groups G.

Since the groups G are Stein manifolds and all spherical functions are holomorphic on G, it looks natural to develop harmonic analysis directly on G using tools of complex analysis (instead of real analysis on U). We discuss here this possibility and find some new possibilities already in classical situation (see also [6]). It is natural to start the complex analysis on G from an integral Cauchy formula, hoping to have an universal analytical tool for harmonic analysis.

1. Cauchy integral formula on symmetric Stein manifolds

The objects of classical harmonic analysis are complex semisimple Lie groups but it is natural to work with more general objects – *symmetric*

Stein manifolds with complex groups of automorphisms. They have the form $Z = G/H$ where G is a simply connected complex semisimple Lie group and H is its involutive subgroup corresponding to a holomorphic involution. The case of a complex semisimple Lie group G_1 is the special case corresponding to $G = G_1 \times G_1$ and a diagonal subgroup $H \cong G_1$. Let A be a maximal Abelian subgroup transversal to H and N be a corresponding unipotent subgroup such that $G^0 = HAN$ is a Zariski open set in G (complex Iwasawa decomposition); let Z^0 be the corresponding dense part of Z. Let $n = \dim Z$, $l = \dim A$ (the rank of Z). Let M be the centralizer of A in H. Then $F = G/AMN$ is a flag manifold. We call the homogeneous manifold $\Xi = G/MN$ *the horospherical manifold.* We have $\dim \Xi = n$. There is a natural fibering $\Xi \to F$ with the fibers A and a natural double fibering $Z \leftarrow G \to \Xi$. Correspondingly points $\zeta \in \Xi$ parameterize some submanifolds in Z which are called *horospheres* $E(\zeta)$ (projections on Z of MN-classes in G). They have codimension l. We have a dual family of submanifolds $S(z), z \in Z$, on Ξ, also of codimension l; we call them *pseudospheres.*

Let us take characters $\delta(a)$ of A and using the Iwasawa decomposition extend them to Z^0. Let us take those that holomorphically extend to all of Z and let $\Delta_j(z|\zeta_0)$, $1 \le j \le l$, be generators of this ring (let them correspond to $\delta_j(a)$). The corresponding characters can be connected with the generators of the lattice of weights dual to non multiple restricted roots. We connect these holomorphic functions with the "initial" horosphere $E(\zeta_0)$ corresponding to the unit element of A: they are constant on horospheres parallel to $E(\zeta_0)$. Acting by G we extend them up to holomorphic functions

$$\Delta_j(z|\zeta)$$

on $Z \times \Xi$. We call these functions *Sylvester's functions:* in the case when Z is the manifold of non degenerate symmetric matrices these functions are the principal minors. In the general case Z^0 is defined by the conditions $\Delta_j(z|\zeta) \ne 0$, a version of the classical Sylvester conditions. The horosphere $E(\zeta)$ is defined by the equations $\Delta_j(z|\zeta) = 1$. Holomorphic extensions of characters on Z have the form $\Delta^\mu(z|\zeta) = \prod \Delta_j(z|\zeta)^{\mu_j}, \mu \in \mathbb{Z}_+^l$. Let $\delta^\mu(a)$ be the corresponding characters on A.

Let us consider differential operators with constant coefficients in logarithmic variables on A. They are defined by polynomial symbols: $P(D)\delta^\mu(a) = P(\mu)\delta^\mu(a)$. We transfer their action to functions $f(z)$ constant on orbits of N - parallel horospheres (denote it as $P(D_z|\zeta_0)$ and $P(D_z|\zeta)$ for other choices of unipotent subgroups).

Now we can write down the formula for *the Cauchy kernel*

$$K(w, z|\zeta) = W(D_z|\zeta)\left(\frac{1}{\prod_{1 \le j \le l}(\Delta_j(w|\zeta) - \Delta_j(z|\zeta))}\right), \quad z, w \in Z, \zeta \in \Xi.$$

We apply here (for fixed ζ) the differential operator of order $n - l$ with the symbol

$$W(\mu) = \prod_{\alpha \in \Sigma_+} \frac{\langle \mu + \rho, \alpha \rangle}{\langle \rho, \alpha \rangle}$$

which is Weyl's polynomial for dimensions of irreducible representations, Σ_+ is a system of restricted roots and ρ is their half-sum. Its appearance is surprising since this formula is a result of purely analytic computations without any appeal to representations.

The kernel has unusual structure. If we apply a differential operator we obtain a combination of fractions whose denominators are monomials in $(\Delta_j(w|\zeta) - \Delta_j(z|\zeta))$ of degree $n - l + 1$ with relatively complicated coefficients. Their choice is defined by the condition that the integrand in the Cauchy formula must be a closed form. The representation with the differential operator is more compact and explicit and I do not think that it would be possible to obtain the applications below using an expression for the kernel as a sum. Of course for $l = 1$ there is only one factor and one term in the sum, so there is no difference in the 2 versions. It is remarkable that the operator with Weyl's symbols delivers the closed form.

We need also 2 remarkable differential forms. Let $\omega(z; dz)$ be the invariant holomorphic n-form on Z and $\lambda(z|\zeta; d\zeta)$ be the similar $(n-l)$-form on pseudospheres $S(z), z \in Z$, of Ξ which is holomorphic on z. Then *we have the Cauchy formula on* Z:

$$\int_\Gamma K(w, z|\zeta)f(z)\omega(z; dz) \wedge \lambda(z|\zeta; d\zeta) = c(\Gamma)f(w), \quad f \in \mathcal{O}(Z).$$

Here Γ is a cycle in $Z \times \Xi$ outside of the singularities of K. The integrand is the closed form and the coefficient $c(\Gamma)$ depends on the cohomology class of Γ (it has an explicit representation). It is crucial that in the Cauchy formula on symmetric Stein manifolds the codimension of cycles equal the rank.

2. The Cauchy-Fantappie formula for cycles of codimensions higher than 1

To obtain this Cauchy formula we need to develop some new tools in multidimensional complex analysis which are, probably, of broader interest. Leray [12, 13] found the universal structure of multidimensional analogues of Cauchy formula – the Cauchy-Fantappie formula. We want to construct an analogue of the Cauchy-Fantappie formula for cycles of higher codimensions. As in Leray's case it is important to find an universal closed form convenient for specific computations. We suggest the form

$$\det\left[\xi_1, \xi_2, \ldots, \xi_l, (d\xi_1 \partial/\partial p_1 + \cdots + d\xi_l \partial/\partial p_l)^{\{n-l\}}\right] \frac{1}{p_1 \cdots p_l}\bigg|_{p_j = \langle \xi_j, w-z \rangle}$$

$$\cdot f(z) \wedge dz_1 \wedge \cdots \wedge dz_n.$$

Here $f(z)$ is a holomorphic function in a domain in $\mathbb{C}^n$; columns ξ_j are n-vectors, the last column repeats $n - l$ times, and we use the exterior product of 1-forms in computing the determinant (for this reason it is not zero). The determinant gives a differential operator on p of order $n - l$. After its application we make the substitution. The result is a holomorphic $(2n-l)$-form which is closed (this is not evident). For $l = 1$ we have Leray's form. In the general case the application of the operator gives a sum of negative monomials. Their coefficients can be directly computed from the condition of closedness (similar computations have appeared in other problems) but for our applications the special representation with a differential operator and a determinant is crucial.

Different specializations of the formula, roughly speaking, correspond to different choices of ξ_j as functions of Z on different cycles in the holomorphy domain of f avoiding the singularities (this can be very non trivial). It is interesting that we need basically all modifications of the Cauchy-Fantappie formula which were considered by Leray. For example, for the Cauchy formula in Ξ we need an analogue of Leray's construction where the function is reconstructed through a differential operator of boundary values. Also extremely important for us is an explicit form of the 2nd Cauchy-Fantappie formula which was found by Henkin and me [8] and its connection with the inversion of the Penrose transform which I found this year [7].

3. Holomorphic horospherical duality for (Z, Ξ)

The first step in the construction of the complex analysis on Z corresponds to the classical harmonic analysis (finite dimensional representations). The central result is

The spaces of holomorphic functions $\mathcal{O}(Z)$ and $\mathcal{O}(\Xi)$ are isomorphic as G-modules.

On the algebraic level this result is known but in this analytic form it is surprising, both from from point of view of Lie groups and of complex analysis: we have an equivalency of reducible non unitary representations on non isomorphic homogeneous G-spaces, which are not biholomorphically equivalent (Ξ is not Stein and has a singular Stein extension). It is crucial that it is not an abstract isomorphism but that there are remarkable explicit intertwining operators. For $f \in \mathcal{O}(Z)$ we define a explicit holomorphic horospherical transform

$$\hat{f}(\zeta) = \int_{\gamma(\zeta)} \frac{f(z)}{\prod_{1 \leq j \leq l} (1 - \Delta_j(z|\zeta))} \omega(z; dz), \zeta \in \Xi.$$

Here γ is a cycle outside of the singularities of the kernel. It turns out that it is possible to take as such cycles some compact forms of Z (this involves non trivially the geometry of Z).

The inverse operator has the form

$$\check{F}(z) = c \int_{\sigma(z)} W(D)F(\zeta)\lambda(z|\zeta; d\zeta), \quad F \in \mathcal{O}(\Xi),$$

where $\sigma(z)$ is a real form of the pseudosphere $S(Z)$.

The proof that these operators are inverse is a direct corollary of the Cauchy formula. We need to take a cycle Γ with separated variables: it must be fibered over the cycle $\sigma(w) \subset S(w) \subset \Xi$ by cycles $\gamma(\zeta)$, whose projection on Ξ has the minimal dimension. The existence of such cycles is a non trivial geometrical fact.

These inversion formulas resemble Radon's inversion formula but in the holomorphic setting they include completely new elements. The analytic duality follows the geometrical duality of horospheres and pseudospheres on Z and Ξ. It is reminiscent of Martineau's duality [14,15], corresponding to the classical projective duality and the Penrose transform. However, in these constructions there are no dualities between spaces of holomorphic functions (rather, spaces of cohomology and holomorphic functions). I do not know other examples of singular integral operators between holomorphic functions which appear in this project. I believe they deserve consideration in a broader context.

The classical results of harmonic analysis are simple corollaries of this duality [6]. On Ξ there is an action of A by "right multiplications" commuting with "left multiplications" by G. Decomposing $\mathcal{O}(\Xi)$ into invariant subspaces relative to this action of A, we obtain irreducible G-

modules (this reduces to analysis for an Abelian compact group – maximal torus in A). The dual subspaces in $\mathcal{O}(Z)$ are irreducible spherical representations. Probably, it is possible to obtain our holomorphic duality starting from irreducible representations, but I do not see a natural way which does not include too much functional analysis.

4. Duality for compact symmetric spaces

The ultimate steps of this project are constructions of dual objects on the horospherical manifold Ξ to semisimple symmetric spaces - real forms of the Stein symmetric manifold Z. Let us start from a compact form X of Z. Let $\widehat{X} \subset \Xi$ be the set of all ζ such that the horosphere $E(\zeta)$ does not intersect X. A priori this set could be empty but it turns out that it is a domain which also can be characterized as the set of all ζ such that $|\Delta_j(z|\zeta)| < 1$ for all j and $z \in X$. It is natural to interpret the domain $\widehat{X}$ as *a geometrical dual object for the compact symmetric manifold X*. This is supported by the next fact on analytic duality:

There is a canonical G-isomorphism of the space $Hyp(X)$ of hyperfunctions on X and the space of holomorphic functions $\mathcal{O}(\widehat{X})$.

The intertwining operators realizing this isomorphism extend the holomorphic isomorphism. Hyperfunctions Φ are functionals on holomorphic functions in a neighborhood of the compact X. Their horospherical transforms $\widehat{\Phi}(\zeta)$ are values on kernels of the horospherical transform as functions of z. To construct the inverse operator we need to extend a functional from these kernels to all holomorphic functions in neighborhoods of X using the Cauchy formula. So it is again a direct consequence of the Cauchy formula. Let us remark that this duality is already new for the sphere S^n. There it turns out that the sphere has a geometrical dual object which is a complex manifold [4, 5].

Let us compare this duality with the unitary trick of H. Weyl. If we decompose $\mathcal{O}(\widehat{X})$ relative to the right action of the compact form of A we obtain the same subspaces of holomorphic functions as for Ξ and correspondingly the same spherical irreducible representations on X. The analytic nature of this coincidence is that X is Riemannian and invariant differential operators are elliptic. Therefore all their eigenfunctions (in which irreducible representations are realized) are holomorphically extended to a complex neighborhood of X, which in this case coincides with all of Z. This is the reason why, on the level of irreducible representations, there is no difference between complex and compact groups. But in the more broad analytic picture the difference is essential and we operate instead on the set of irreducible representations with other geometrical dual objects.

5. Duality for Riemannian symmetric space of noncompact type

Let us connsider $X = G_{\mathbb{R}}/H_{\mathbb{R}}$, another real form of Z which is Riemannian of noncompact type. Here we have real forms of G, H such that $H_{\mathbb{R}}$ is maximal compact in $G_{\mathbb{R}}$. In the usual harmonic analysis on X, $L^2(X)$ is considered but we will work with holomorphic functions. The manifold X has a canonical Stein neighborhood which was defined in [1] and which I suggested to call *the complex crown of X*: $X \subset \text{Crown}(X) \subset Z$. We consider the $G_{\mathbb{R}}$-module $\mathcal{O}(\text{Crown}(X))$. The crucial geometrical fact is that the crown is convex relative to horospheres: the complement is the union of some horospheres $E(\zeta)$. This is a reformulation of the result [9]. *As a dual object we take the manifold $\widehat{X}$ of all ζ such that the horospheres $E(\zeta)$ intersect the crown.* We define an analogue of the Penrose transform

$$\mathcal{P} : H^{(n-l)}(\widehat{X}, \mathcal{O}) \to \mathcal{O}(\text{Crown}(X)).$$

We just integrate forms $\phi \wedge \mu(z(\zeta, d\zeta))$, where ϕ is a $\bar{\partial}$-closed $(n-l)$-form, on pseudospheres $S(z)$. Technically it is convenient to work with a holomorphic envelope of Ξ where the $S(z)$ are compactified. It is possible to construct also an explicit operator

$$\mathcal{Q} : \mathcal{O}(\text{Crown}(X)) \to H^{(n-l)}(\widehat{X}, \mathcal{O})$$

such that *the operator $\mathcal{P}\mathcal{Q}$ is the unit operator. This implies that the Penrose transform $\mathcal{P}$ is surjective.*

The operator $\mathcal{P}$ has a relatively small kernel with an explicit description. The construction of the operator $\mathcal{Q}$ follows the concept of the holomorphic language for analytic cohomology [2] and uses the Cauchy formula on Z in the form similar to the 2nd Cauchy-Fantappie formula (cf. [8]).

If we translate to irreducible representations, this duality corresponds to Helgasson's conjecture which was proved in [11]. Eigenspaces of invariant differential operators on X correspond in the dual picture to $A_{\mathbb{R}}$-invariant cohomology on $\widehat{X}$ which are equivalent to cohomology on $F \setminus F_{\mathbb{R}}$ (the complement in the manifold of complex flags of the submanifold of real ones) with the coefficients in some line bundles. Then the operator $\mathcal{P}$ corresponds to the Poisson transform and $\mathcal{Q}$ to the operator of hyperfunction- boundary values. Only for a discrete set of eigenvalues does the Poisson transform have a kernel.

It was found in [10] that all causal symmetric manifolds Y can be realized as an edge of the boundary of $\text{Crown}(X)$ for some X. In this way we have an interpretation of one multiplicity of maximal continuous spectrum for such Y, free of L^2-restrictions.

The final aim of this project is *the duality for pseudo Riemannian semisimple symmetric manifolds* $X = G_{\mathbb{R}}/H_{\mathbb{R}} \subset Z$. If $H_{\mathbb{R}}$ is not compact then there are some remarkable $G_{\mathbb{R}}$-invariant tubes in Z with the edge X which are, as a rule, non Stein.

The aim is to find appropriate geometrical dual objects for them in Ξ *such that an analytic duality connects some cohomology in dual manifolds.*

In a sense this is a refinement of the old program [3]. Some components of the final picture have already been constructed but we will not discuss them here.

References

[1] D. N. AKHIEZER and S. G. GINDIKIN, *On Stein extensions of real symmetric spaces*, Math. Ann. **286** (1990), 1–12.

[2] M. G. EASTWOOD, S. G. GINDIKIN and H.-W. WONG, *Holomorphic realization of $\bar{\partial}$-cohomology and constructions of representations*, Jour. Geom. Phys. **17** (1995), 231–244.

[3] I. GELFAND and S. GINDIKIN, *Complex manifolds whose skeletons are real semisimple Lie groups and holomorphic discrete series*, Functional Anal. Appl. **11** (1977), 19–27.

[4] S. GINDIKIN, *Complex horospherical transform on real sphere*, In: "Geometric Analysis of PDE and Several Complex Variables", Contemporary Mathematics **368**, American Math. Soc., 2005, 227–232.

[5] S. GINDIKIN, *Holomorphic horospherical duality "sphere-cone"*, Indagationes Mathematicae N.S. **16** (2005), 487–497.

[6] S. GINDIKIN, *Harmonic analysis on symmetric manifolds from the point of view o complex analysis*, Japanese J. Math. **1** (2006), 87–105.

[7] S. GINDIKIN, *Second Cauchy-Fantappie formula and Matineu duality*, J. of Algebra (2007), to appear.

[8] S. GINDIKIN and G. HENKIN, *The Cauchy-Fantappie on projective space*, Amer. Math. Soc. Transl. **146** (1990), 23–32.

[9] S. GINDIKIN and B. KRÖTZ, *Invariant Stein domains in Stein symmetric spaces and a non-linear complex convexity theorem*, Intern. Math. Res. Not. **18** (2002), 959–971.

[10] S. GINDIKIN and B. KRÖTZ, *Complex crowns of Riemannian symmetric spaces and non-compactly causal symmetric spaces*, Transactions AMS **354** (2002), 3299–3327.

[11] M. KASHIWARA, A. KOWATA, K. MINEMURA, K. OKAMOTO, T. OSHIMA and M. TANAKA, *Eigenfunctions of invariant differential operators on a symmetric space*, Ann. Math. **107** (1978), 1–39.

[12] J. LERAY, *Function de variables complexe: sa représentation comme somme de puissances négatives de functions linéaires*, Atti. Accad. Naz. LinceiRend. Cl. Sci. Fis. Mat. Nat. **20** (1956), 589–590.

[13] J. LERAY, *Le calcul differentiel et integral sur une variété analytique complexe (Problème de Cauchy, III)*, Bull. Soc. Math. France **87** (1959), 81–180.

[14] A. MARTINEAU, *Indicatrices des fonctionellea analitiques at inversionde la transformation de Roiurier-Borel pat la transformation de la Laplace*, C.R. Acad. Sci. **255** (1962), 2888–2890.

[15] A. MARTINEAU, *Sur la topologie des espaces de functions holomorrphes*, Math. Ann. **163** (1966), 62–88.

Historical reminiscences on the Gauss class number problem

Dorian Goldfeld

Consider an imaginary quadratic field $\mathbb{Q}(\sqrt{D})$ with $D < 0$ where D is a rational integer which is either $\equiv 1 \pmod 4$ or of the form $D = 4m$ with $m \equiv 2$ or $3 \pmod 4$.

Define

$$h(D) = \#\left\{ \frac{\text{group of nonzero fractional ideals } \frac{a}{b}}{\text{group of principal ideals } (\alpha), \ \alpha \in k^\times} \right\}$$

to be the cardinality of the ideal class group of k. In [4] it was shown (using the language of binary quadratic forms) that $h(D)$ is always finite. Gauss conjectured that

$$h(D) \to \infty \quad \text{as} \quad D \to -\infty,$$

which was first proved in [11].

The Disquisitiones contains tables of binary quadratic forms with small class numbers. Gauss also conjectured that his tables were complete which indicated that he believed that there should exist an effective algorithm to determine all imaginary quadratic fields with a given class number. The question of whether such an algorithm exists or not is now called the Gauss class number problem. The case of class number one is particularly interesting, because in this case, it can be shown that the imaginary quadratic field has the unique factorization property – *that every integer in the field can be uniquely factored into primes.* For the case of class number one, Gauss' conjecture takes the explicit form that there

The author's research is supported in part by NSF grant DMS 0354582.

are only nine discriminants

$$D = -3, -4, -7, -8, -11, -19, -43, -67, -163$$

where $\mathbb{Q}(\sqrt{D})$ has class number one. The class number one problem was first solved by Heegner (see [10]), but his proof was not generally recognized by the mathematical community until the work of Stark [15]. Later Stark [16] and Baker [1] gave complete solutions to the class number one and two problems.

My own interest in imaginary quadratic fields was sparked when I first met Chowla at the Institute for Advanced Study in Princeton in 1973. Chowla loved to show how easy it was to compute the quadratic residue symbol by hand and he was always talking about the seemingly impossible to solve problems (such as the class number problem) which arose in such a simple and elegant manner. For him, the class number problem was the quintessential math problem. There was nothing more beautiful. Chowla and I spent more and more time together and ultimately wrote several papers together. But what I remember most is that he instilled in me the love of quadratic fields.

The year I was at the Institute for Advanced Study was the same year that Enrico Bombieri was also a visiting member. I had many mathematical discussions with Enrico. I distinctly remember how he would talk for hours, telling me his innermost thoughts on any question I asked. Never before did I have an opportunity to have such a close association with one of the greatest mathematicians of our time. As an aspiring analytic number theorist I knew then and there that I had to go to Italy to study with Bombieri. In 1974 Bombieri moved to the Scuola Normale Superiore in Pisa and I followed him and stayed in Pisa for two years.

For me, Pisa was an ideal city to live in. I bicycled every day past the Leaning Tower to the Scuola Normale. I had no teaching duties, my life was free to pursue only mathematics. I began to think of Chowla and the class number problem of Gauss. The only approach known at that time for attacking the class number problem was based on an idea of Hugh Montgomery (see [14]).

The idea was that if the Riemann zeta function or a fixed Dirichlet L-function had two zeros on the line $\Re(s) = \frac{1}{2}$ which were too close together (of if the zeta function had a double zero) then one could solve the class number problem. I had heard Montgomery lecture on this approach and quickly came up with a simple proof of my own. I realized that the method would work, not only for Dirichlet L-functions, but for any L-function which satisfied certain conditions. The conditions required were the following:

- *An L-function $L(s) = \sum_{n=1}^{\infty} a(n)n^{-s}$ (with real coefficients $a(N)$, $n = 1, 2, 3 \ldots$) where the series converges absolutely in some half-plane $\Re(s) \gg 1$.*

- *$L(s)$ has a meromorphic continuation in $s \in \mathbb{C}$ with at most finitely many poles.*

- *$L(s)$ satisfies a functional equation of the type*

$$A^s G(s)L(s) = \pm A^{1-s} G(1-s)L(1-s)$$

where $A > 0$ and $G(s)$ is a product of Gamma functions.

- *$L(s)$ has a representation as an Euler product:*

$$L(s) = \prod_{i=1}^{k} \prod_{p} (1 - \alpha_{p,i}\, p^{-s})^{-1},$$

where k is a fixed integer ≥ 1, $\alpha_{p,i} \in \mathbb{C}$, and the product goes over all rational primes p.

I was then able to show that if an L-function of the above type existed and it had a zero of sufficiently high multiplicity on the line $\Re(s) = \frac{1}{2}$ then the class number problem of Gauss could be solved. The only problem with the approach was that at the time there did not seem to be any candidate L-functions with multiple zeros on the critical line. This seemed to be just another good idea that wasn't going anywhere. Montgomery, on the other hand, pushed his original idea in another direction and went on to make his now famous pair correlation conjecture [13] about the distribution of zeros of L-functions in short intervals which ultimately led to the introduction of random matrix models into number theory.

Since my ideas for the class number problem seemed dead ended, I applied the methods I had developed to a related problem. It had been known since the work of Hecke-Landau (see [7]) if an imaginary quadratic field existed with large discriminant and small class number then the generalised Riemann hypothesis was false and the standard Dirichlet L-function associated to the imaginary quadratic field had to have a real zero near $s = 1$. Such a zero is commonly termed a Siegel zero. I was able to show that in this hypothetical situation it was possible to obtain an asymptotic formula for the precise location of the elusive Siegel zero. I showed this to Bombieri and he asked that I submit the paper to the Annali di Pisa where it was later published. When Schinzel (much later) visited Bombieri in Pisa we worked together and extended the theorem (see [8]) to real quadratic fields.

In 1974, after finding the asymptotic formula for the Siegel zero I began again to try to show that the Siegel zero cannot exist which would solve the class number problem of Gauss. It was only necessary to find an auxilliary L-function which could have a multiple zero on the critical line. I began to spend a lot of time in the library of the Scuola Normale looking for papers connected with zeros of L-functions of any type. Then one day I stumbled on the paper of [2] which discussed L-functions associated to elliptic curves (a subject which at the time I knew very little about). In fact, elliptic curves were not really an object of study by most analytic number theorists in 1974, but luckily for me, Bombieri was an exception. I became obsessed with elliptic curves and with the help of Bombieri started to master the analytic side of the theory. The striking idea which caught my initial attention was the conjecture of Birch and Swinnerton-Dyer which says that the L-function of an elliptic curve over $\mathbb{Q}$ has a zero of multiplicity r at the center of the critical strip where r is the rank of the Mordell-Weil group of the elliptic curve. Birch and Swinnerton-Dyer had many examples of elliptic curves with rank $r = 2$ in their original paper. Here, finally, was an open door to the possible solution of the class number problem!

I immediately went to Bombieri and asked him if he thought it might be possible to use the L-functions of elliptic curves to solve the class number problem. His response was that although he felt very strongly that elliptic curves (over $\mathbb{Q}$) should exist with high rank (and a priori that L-functions should exist having zeros of high multiplicity at the center of the critical strip) he was very skeptical that the methods originally introduced by Montgomery would work in this situation. I kept going back to him telling him that if an elliptic curve L-function had a zero of multiplicity 3 at the center of the critical strip then I could solve the class number problem. His response was that I should write up the proof and he would look at it carefully. This I did rather quickly since I had already worked much of the proof in the first part of my stay in Pisa. Bombieri studied my work for several weeks and when he asked me to meet again, I was quite nervous while waiting at the door. His response was very positive and extremely encouraging, but he wanted to see every detail worked out explicitly. I came back a few weeks later with a manuscript that was 3 times as long and after carefully studying it he agreed that my proof was correct. I had shown [5] that Gauss' class number problem could be solved if there existed one elliptic curve (over $\mathbb{Q}$, satisfying a ramification condition) whose L-function had holomorphic continuation and had a triple (or higher order) zero at the center of the critical strip.

Birch and Swinnerton-Dyer had already demonstrated that there exist L-functions associated to elliptic curves with complex multiplication that had holomorphic continuation and double zeroes at the center of the critical strip. Try as I might, however, I could not improve my theorem to this situation. It seemed vitally necessary to have the existence of a triple zero. A double zero was just not good enough. When Schinzel visited Pisa, he pointed out to me a paper of [19] where it was shown that there exists an elliptic curve over $\mathbb{Q}$ with complex multiplication by i which had rank 4. Since L-functions associated to complex multiplication elliptic curves (over $\mathbb{Q}$) were known to be entire, it was not necessary to assume the Shimura-Taniyama-Weil conjecture (now proved completely in ([17], [3]). It was only necessary to establish the Birch-Swinnerton-Dyer conjecture for one specific complex multiplication curve of rank three or more. By Wiman (see [19]), we knew such curves existed.

I had to wait another 10 years before the existence of one example of an entire L-function of an elliptic curve with a triple zero at the center of the critical strip was established in [9]. This example led to the definitive solution of the Gauss class number problem. The complete story is told in [6] and then again in [12], [7].

Let me conclude with the description of an interesting research problem. Let $\mathbb{Q}(\sqrt{D})$ denote an imaginary quadratic field with class number $h(D)$ as in the beginning of these reminiscences. If there exists a suitable elliptic curve (over $\mathbb{Q}$) whose L-function has a zero of multiplicity r at the center of the critical strip then one may obtain the effective bound

$$h(D) \gg |D|^{r-2-\epsilon} \tag{1.1}$$

for any fixed $\epsilon > 0$ and all $D \to -\infty$. Now, it is known, (see [18]) that in the case of elliptic curves over function fields that the rank of the elliptic curve can be very high. It is possible that elliptic curves of conductor N exist (over $\mathbb{Q}$) with ranks as large as

$$\frac{\log N}{\log \log N}.$$

The question arises, is it possible to modify the proof of (1.1) in the case where r depends on $|D|$ and is assumed to be as large as $\frac{\log |D|}{\log \log |D|}$? What type of lower bound for $h(D)$ would one obtain in this situation? Of course, if the bound one gets is too strong, this would give a way of showing that very high rank curves should not exist over $\mathbb{Q}$ (modulo the Birch-Swinnerton-Dyer conjecture).

References

[1] A. BAKER, *Imaginary quadratic fields with class number "2"*, Annals of Math. **94** (1971), 139–152.

[2] B. J. BIRCH and H. P. F. SWINNERTON-DYER, *Notes on elliptic curves. II*, J. Reine Angew. Math. **218** (1965), 79–108.

[3] C. BREUIL, B. CONRAD, F. DIAMOND and R. TAYLOR, *On the modularity of elliptic curves over Q: wild 3-adic exercises*, Journal of the American Mathematical Society **14** (2001), 843–939.

[4] C. F. GAUSS, "Disquisitiones Arithmeticae", Göttingen, 1801; English translation by A. Clarke, revised by W. Waterhouse, New Haven, Yale University Press, 1966; reprinted by Springer Verlag, 1986.

[5] D. GOLDFELD, *The class number of quadratic fields and the conjecture of Birch and Swinnerton-Dyer*, Ann. Scuola Norm. Sup. Pisa Cl. Sci. (4) **3** (1976), 624–663.

[6] D. GOLDFELD, *Gauss' class number problem for imaginary quadratic fields*, Bull. Amer. Math. Soc. **13** (1985), 23–37.

[7] D. GOLDFELD, "Automorphic Forms and L-Functions for the Group $GL(n, R)$", Cambridge Studies in Advanced Mathematics, n. 99, Cambridge University Press, 2006.

[8] D. M. GOLDFELD and A. SCHINZEL, *On Siegel's zero*, Ann. Scuola Norm. Sup. Pisa Cl. Sci. (4) **2** (1975), 571–583.

[9] B. GROSS and D. B. ZAGIER, *Heegner points and derivatives of L-series*, Invent. Math. **84** (1986), 225–320.

[10] K. HEEGNER, *Diophantine Analysis und Modulfunktionen*, Math. Z. **56** (1952), 227–253.

[11] H. HEILBRONN, *On the class number in imaginary quadratic fields*, Quarterly J. of Math. **5** (1934), 150–160.

[12] H. IWANIEC and E. KOWALSKI, *Analytic Number Theory*, AMS Colloq. Publications **53** (2004).

[13] H. MONTGOMERY, *The pair correlation of zeros of the zeta function*, Analytic Number Theory (Proceedings of Symposia in Pure Mathematics) **24** (1973), 181-193.

[14] H. MONTGOMERY and P. WEINBERGER, *Notes on Small Class Numbers*, Acta. Arith. **24** (1974), 529-542.

[15] H. STARK, *A complete determination of the complex quadratic fields of class number one*, Mich. Math. J. **14** (1967), 1–27.

[16] H. STARK, *A transcendence theorem for class number problems I, II*, Annals of Math. **94** (1971), 153–173; **96** (1972), 174–209.

[17] R. TAYLOR and A. WILES, *Ring-theoretic properties of certain Hecke algebras*, Ann. of Math. **141** (1995), 553–572.

[18] D. ULMER, "Elliptic curves and analogies between number fields and function fields", Heegner points and Rankin L-series, Math. Sci. Res. Inst. Publ. Vol. 49, Cambridge Univ. Press, Cambridge, 2004, 285–315.

[19] A. WIMAN, *Über rational Punkte auf Kurven* $y^2 = x(x^2 - e^2)$, Acta Math. **77** (1945), 281-320.

From $2^{\sqrt{2}}$ to polarizations on abelian varieties

David Masser

Abstract. The first big success of the general method of Gelfond (who would have been 100 years old next year) was in proving the transcendence of numbers as in the title, as conjectured by Hilbert in his Seventh Problem. This was vastly generalized by Baker in his theory of linear forms in logarithms. The elliptic analogues of the theory can be extended in order to estimate isogenies between elliptic curves. Similarly for the abelian analogues; but here the theory can be taken a step further to estimate polarizations on a single abelian variety. In our talk we give an account of these developments and we mention some open problems.

1. Hilbert in the 1900 International Mathematical Congress posed the problem of proving the transcendence of α^{β} for algebraic α, β; for example $2^{\sqrt{2}}$ or $e^{\pi} = i^{-2i}$ or the well-known "near-miss"

$$e^{\pi\sqrt{163}} = 262537412640768743.9999999999992500725\ldots$$

Gelfond did this for e^{π} in 1929 and Kuzmin did it for $2^{\sqrt{2}}$ in 1930. The general result was established in 1934 by Gelfond and Schneider independently: take $\alpha \neq 0$ algebraic, take $\log\alpha$ in $\mathbf{C}$, and take β algebraic irrational. Then $\alpha^{\beta} = \exp(\beta\log\alpha)$ is transcendental. Here is an absurdly concise summary of the proof in four steps.

(P1): If not, then the functions e^z, $e^{\beta z}$ take algebraic values at $z = \log\alpha$ and so do their derivatives, even at all integer multiples of $\log\alpha$.

(P2): With arithmetic, construct a non-zero polynomial $P(X, Y)$ with rational integer coefficients such that $f(z) = P(e^z, e^{\beta z})$ has high order zeros at many of these integer multiples.

(P3): With analysis, deduce that $f(z)$ is identically zero; that is, e^z, $e^{\beta z}$ are algebraically dependent.

(P4): With algebra, conclude that β must be rational; a contradiction.

Equivalently: take $u_1, u_2 \neq 0$ in $\mathbf{C}$ with e^{u_1}, e^{u_2} algebraic. Then if $\frac{u_1}{u_2}$ is algebraic, it is rational.

In this form it was generalized by Baker in 1966 to $u_1, \ldots, u_n$; now the linear dependence of these numbers over the field of all algebraic numbers implies their linear dependence over $\mathbf{Q}$. This gives for example the transcendence of $2^{\sqrt{2}} e^{\pi}$. That of $2^{\sqrt{2}} + e^{\pi}$ is still unknown, and even the irrationality too (and by the way also the irrationality of $e + \pi$).

2. The elliptic analogues were formulated almost immediately. The e^z above gives an analytic map from $\mathbf{C}$ to $\mathbf{C} \setminus \{0\}$ with period group $2\pi i \mathbf{Z}$. Much larger period groups are possible. In fact take any lattice $\Omega = \mathbf{Z}\omega_1 + \mathbf{Z}\omega_2 \subset \mathbf{C}$. Then there is a meromorphic map $\wp(z)$ from $\mathbf{C}$ to $\mathbf{C} \cup \{\infty\}$ with period group Ω: the Weierstrass elliptic function.

We used derivatives in (P1) above, and there is the famous differential equation

$$(\wp'(z))^2 = 4(\wp(z))^3 - g_2 \wp(z) - g_3$$

with g_2, g_3 depending only on Ω. So we need special Ω having algebraic g_2, g_3. This is easy to secure. Interesting examples are $\omega(\mathbf{Z} + \mathbf{Z}\sqrt{-1})$ (square) and $\omega^*(\mathbf{Z} + \mathbf{Z}\frac{-1+\sqrt{-3}}{2})$ (hexagonal or triangular) with

$$\omega = \frac{\Gamma\left(\frac{1}{4}\right)^2}{2\sqrt{2\pi}}, \quad g_2 = 4, \ g_3 = 0$$

and

$$\omega^* = \frac{\Gamma\left(\frac{1}{3}\right)^3}{2^{4/3}\pi}, \quad g_2 = 0, \ g_3 = 4$$

respectively.

The elliptic analogue of the Gelfond-Schneider theorem is due to Schneider in 1936; now $\wp(u_1), \wp(u_2)$ are algebraic or infinite. The analogue of Baker's theorem for a single elliptic function was proved by me in 1975 in the special so-called complex multiplication case, and in general by Bertrand and me in 1980. In fact Schneider treated any two $\wp_1, \wp_2$ with $\wp_1(u_1), \wp_2(u_2)$ algebraic or infinite.

3. As an example, let us try to prove that $\frac{\omega^*}{\omega}$ above is transcendental; here $\wp(\omega) = \wp^*(\omega^*) = \infty$. Use (P1) with $\wp(z)$, $\wp^*(\frac{\omega^*}{\omega}z)$; the conclusion in (P3) is that these functions are algebraically dependent. This no longer implies that $\frac{\omega^*}{\omega}$ is rational, though.

To see what does follow, it is better to use the language of elliptic curves (which are special cases of commutative algebraic groups). Generally

$$y^2 = 4x^3 - g_2x - g_3 \qquad (E)$$

defines an elliptic curve E, and $P = (\wp, \wp')$ gives an isomorphism between $\mathbf{C}/\Omega$ and the set of complex points $E(\mathbf{C})$. Similarly P^* and E^*; and now $f = 0$ in (P3) defines an algebraic subvariety $V \subset E \times E^*$ (if we use $z^* = \frac{\omega^*}{\omega}z$ for the second variable). The image T of V in $(\mathbf{C}/\Omega) \times (\mathbf{C}/\Omega^*)$ contains $\mathbf{C}(1, \frac{\omega^*}{\omega})$; and mapping T back gives a subset Z of V which is no algebraic variety but is an infinite group. So its Zariski-closure $H = \overline{Z}$ is an algebraic group of positive dimension.

We can now exploit the group-theoretical consequences of $H \subset E \times E^*$. There results a sequence of maps

$$E \longrightarrow E/F \longrightarrow E^*/F^* \longrightarrow E^*$$

for finite subgroups $F \subset E$, $F^* \subset E^*$, in which the first is canonical, the middle is an isomorphism, and the last is multiplication by the order of F^*. So we get a surjective $\varphi : E \to E^*$, which is called an isogeny. By means of a translation we can assume that φ takes the identity of E to the identity of E^*.

It is not hard to see that in our example this is impossible. Thus the corresponding map $\beta : \mathbf{C}/\Omega \to \mathbf{C}/\Omega^*$ is well-known to be multiplication by some non-zero complex number, which we could also denote by β. In particular $\beta\Omega \subseteq \Omega^*$. But β cannot take squares into triangles! Thus indeed $\frac{\omega^*}{\omega}$ is transcendental.

4. Here we could determine quite quickly that there is no isogeny: E, E^* are not isogenous. But what about

$$y^2 = x(x+1)(x+4), \quad y^2 = x(x-1)(x-9)$$

for example? We know hardly anything about the period lattices.

It was realized in 1985 by the Chudnovskys that the very same technique of transcendence can sometimes be used to settle this problem. Suppose $\varphi : E \to E^*$ exists, so that $\beta\Omega \subseteq \Omega^*$. Pick fixed $\omega \neq 0$ in Ω; then $\omega^* = \beta\omega$ is in Ω^*. *If now ω^* happens also to be fixed*, then the machinery of (P1), (P2), (P3) above delivers H of bounded "degree". This bound has to be calculated, and one finds (for simplicity assume that g_2, g_3, g_2^*, g_3^* are rational)

$$\deg H \leq c_1 \left(\max\{1, h(E), h(E^*)\} + \log \deg \varphi \right)^\kappa$$

for absolute c_1, κ, where h measures a logarithmic height such as

$$h(E) = \log \max\{|g_2|, |g_3|\}$$

in case g_2, g_3 are in $\mathbf{Z}$. This H leads to a new isogeny φ' with

$$\deg \varphi' \leq 9(\deg H)^2 \leq c_2 \left(\max\{1, h(E), h(E^*)\} + \log \deg \varphi \right)^{2\kappa}.$$

The new degree looks smaller, so we can use induction or better simply assume that $\deg \varphi$ above was already minimal. Thus $\deg \varphi \leq \deg \varphi'$ and it follows easily that

$$\deg \varphi \leq c_3 \left(\max\{1, h(E), h(E^*)\} \right)^{2\kappa}.$$

Algebraically this bounds the degrees of the rational functions defining φ, and now there are established procedures to determine if φ really exists.

But what if $\omega^* = \beta\omega$ above was not fixed? Then we have to write

$$\beta\omega = b_1\omega_1^* + b_2\omega_2^*$$

for fixed basis elements of Ω^*, and so Baker's method has to be used. This and the subsequent developments are due to Wüstholz and the writer.

5. All this generalizes to abelian varieties A instead of elliptic curves E. But now there are no canonical equations like (E) above. Amazingly there is a canonical height $h_F(A)$, due to Faltings. But for the transcendence machinery one needs equations. These come as soon as we know a "polarization" on A. So, for example, if A, A^* are isogenous with fixed or bounded polarizations, then we can prove that there must be an isogeny of degree at most $c_4 \left(\max\{1, h_F(A), h_F(A^*)\} \right)^\lambda$. This makes effective a step in Faltings's proof of the Mordell conjecture.

6. Actually a polarization on A is itself "more or less" an isogeny! Namely $\pi : A \to \widehat{A}$ for the so-called dual $\widehat{A}$ of A. So can't we apply the above argument to this situation and thereby find a bounded polarization?

No, because we don't know a polarization on A to start with!!

7. Return to $\varphi : A \to A^*$. An ingenious trick due to Zarhin consists in noting that the abelian variety

$$Z = \mathcal{Z}(A) = A^4 \times \widehat{A}^4$$

always has a "principal" polarization, the simplest possible.

And from an isogeny $A \to A^*$ we get an isogeny $Z = \mathcal{Z}(A) \to \mathcal{Z}(A^*) = Z^*$ or

$$A^4 \times \widehat{A}^4 \longrightarrow (A^*)^4 \times (\widehat{A^*})^4. \tag{Z}$$

By the above, there is a bounded isogeny (Z). But getting back to a bounded $A \to A^*$ is not so easy. For example, suppose that the second abelian variety A^* happened to be $\widehat{A}$. Then (Z) looks like $A^4 \times \widehat{A}^4 \to \widehat{A}^4 \times A^4$. There now seems nothing to stop this map being just a permutation of the factors, and such a simple map cannot possibly give rise to $A \to \widehat{A}$.

The solution is to find not just one isogeny (Z) but all of them; the whole homomorphism group $\mathrm{Hom}(Z, Z^*)$. This can be done by extending the transcendence techniques. Then out of the whole group one can pick out a bounded $A \to A^*$. So the hypothesis of fixed polarizations can be dropped.

8. Thus indeed we can find a bounded isogeny $\pi : A \to \widehat{A}$. Unfortunately this does not always define a polarization; there must be a certain positivity condition too, with respect to the endomorphism ring $\mathrm{End}(A) = \mathrm{Hom}(A, A)$.

In the simplest case $\mathrm{End}(A) = \mathbf{Z}$ it suffices to change π into $-\pi$; exactly one of these is positive.

In general we don't yet know how to solve this problem. But if $\dim A \le 7$ or if A is simple of arbitrary squarefree dimension we do, because then $\mathrm{End}(A)$ can be described concretely enough. In these cases it follows that A has a polarization of degree at most $c_5 (\max\{1, h_F(A)\})^\mu$. Perhaps this is true even with $\mu = 0$. An obstacle to generalization occurs first at dim $A = 8$, when there is one class of endomorphism ring in Albert's famous classification which defies our analysis.

9. In these positivity situations we have to use the so-called Asymmetric Geometry of Numbers. We illustrate this topic with a lattice $\Lambda \subset \mathbf{R}^2$.

For any $X > 0$, $Y > 0$ with $XY = \det \Lambda$ we can find thanks to Minkowski non-zero (x, y) in Λ with

$$|x| \leq X, \quad |y| \leq Y.$$

But if we want the positivity $x > 0, y > 0$ then we cannot in principle bound them separately. For example $\mathbf{Z}^2$ contains no points at all with

$$0 < x \leq X, \quad 0 < y \leq Y$$

when $X < 1$, however large the value of XY. Chalk proved that in general there is a point with

$$x > 0, \quad y > 0, \quad xy \leq \det \Lambda;$$

with a generalization to higher dimensions.

We also need an extra argument using the following technical estimate. Let K be a totally real number field, and let $Q(x, y, z)$ be a quadratic form in $K[x, y, z]$ with total signature $(+, -, -)$. Then there are ξ, η, ζ in the ring of integers of K such that $q = Q(\xi, \eta, \zeta)$ is totally positive with

$$\mathrm{Norm}_{K/\mathbf{Q}}(q) \leq 2^{2[K:\mathbf{Q}]}|\mathrm{disc}(K)|^5|\mathrm{Norm}_{K/\mathbf{Q}}(\mathrm{disc}(Q))|^{1/3} \qquad (N)$$

for certain discriminants. The case $K = \mathbf{Q}$ is due to Blaney (in n variables), with the upper bound $2^{n-1}|\mathrm{disc}(Q)|^{1/n}$.

10. In fact the set of polarizations on an abelian variety A can be identified with a "positive" subset of the Néron-Severi group NS(A), itself sitting inside $\mathcal{H} = \mathrm{Hom}(A, \widehat{A})$. Chalk's result suffices to estimate a polarization when the endomorphism algebra $\mathbf{Q} \otimes_{\mathbf{Z}} \mathrm{End}(A)$ is a totally real number field. But this is not the only case in Albert's classification. The most difficult case for us turns out to be a totally indefinite quaternion algebra $\mathcal{A}$ over such a field. Now NS(A) is too big, and we can deal efficiently only with a natural complement in $\mathcal{H}$, which for want of a better name we call the Severi-Néron group SN(A). Every polarization gives rise to a Rosati involution on $\mathcal{A}$ which also has a positivity property. There is also a canonical involution; but unfortunately it is not positive so it cannot come from a canonical polarization. It is here that we use the estimate (N) to find a bounded positive involution and thereby a bounded polarization. We have indeed come a long way from $2^{\sqrt{2}}$.

Eigenvalue statistics and lattice points

Zeév Rudnick

Abstract. One of the more challenging problems in spectral theory and mathematical physics today is to understand the statistical distribution of eigenvalues of the Laplacian on a compact manifold. There are now several challenging conjectures about these, originating in the physics literature. In this survey, a version of a talk delivered as the Colloquio De Giorgi at the Scuola Normale Superiore of Pisa in May 2006, I will describe what is conjectured and what is known the very simple case of the flat torus, where the problems amount to counting lattice points in annuli and have a definite arithmetic flavour.

Eigenvalue problems. Let M be a smooth, compact Riemannian manifold, Δ the Laplace-Beltrami operator associated with the metric. We consider the eigenvalue problem

$$\Delta\psi + \lambda\psi = 0, \quad \psi \in L^2(M).$$

As is well known, the spectrum is a discrete set of points, which we denote by $0 = \lambda_0 \leq \lambda_1 \leq \lambda_2 \leq \ldots$, whose only accumulation point is at infinity, and there is an orthonormal basis of $L^2(M)$ consisting of eigenfunctions.

Example 1.1. The circle $M = \mathbb{R}/\mathbb{Z}$ with the standard flat metric: the Laplacian is simply d^2/dx^2, as a basis for the eigenfunctions we may take $\sin 2\pi kx$, $\cos 2\pi kx$ ($k > 0$) with eigenvalues $4\pi^2 k^2$ (each having multiplicity 2), together with the constant function 1 (eigenvalue zero).

Example 1.2. The flat torus $\mathbb{R}^2/\mathbb{Z}^2$: here the Laplacian is $\Delta = \frac{\partial^2}{\partial x^2} + \frac{\partial^2}{\partial y^2}$ and as an orthonormal basis of eigenfunctions we may take the exponentials $\exp(2\pi ik \cdot x)$, $k \in \mathbb{Z}^2$, with eigenvalue $4\pi^2|k|^2$.

Spectral counting functions. We define the following spectral functions: A cumulative count

$$n(x) := \#\{m : \sqrt{\lambda_m} \leq x\}$$

Supported by a grant from the Israel Science Foundation.

and a window count

$$n(x, \delta) := \#\{m : x < \sqrt{\lambda_m} \le x + \delta\}.$$

In the case of the torus $\mathbb{R}^2/\mathbb{Z}^2$, the cumulative counting function

$$n(x) = \#\{k \in \mathbb{Z}^2 : \sqrt{4\pi^2|k|^2} \le x\}$$

is exactly the number of lattice points in a disk of radius $t = x/2\pi$. The spectral function $n(x, \delta)$ is then the number of lattice points in an annulus of width $\delta/2\pi$:

$$n(x, \delta) = \#\{k \in \mathbb{Z}^2 : x < 2\pi|k| < x + \delta\}.$$

Weyl's law. Weyl's law gives the asymptotics of $n(x)$ as $x \to \infty$. In our general context of a compact, d-dimensional manifold, Weyl's law says that

$$n(x) \sim c_d \operatorname{vol}(M)x^d, \quad x \to \infty$$

where c_d is a universal constant depending only on the dimension. For the surface case, $c_2 = 1/4\pi$. Thus in the case of the torus, where the area is 1, Weyl's law says that

$$n(x) \sim \frac{1}{4\pi}x^2$$

which is the area of the circle of radius $x/2\pi$. This is consistent with the simplest intuition that the number of lattice points in such a circle is well approximated by its area.

The remainder term. A famous problem is to bound the remainder term for the number $N(t)$ of lattice points in a disk of radius t. An elementary packing argument, attributed to Gauss, gives

$$N(t) = \pi t^2 + O(t)$$

and the circle problem is to find the best exponent θ for which

$$N(t) = \pi t^2 + O(t^{\theta+\epsilon}), \quad \forall \epsilon > 0.$$

This is a problem with a long history. The first nontrivial result is due to Sierpinski (circa 1905) who showed that one can take $\theta = 2/3$. The current record is due to Huxley (2005) giving $\theta = 131/208 = 0.629\ldots$ The well-known conjecture of G. H. Hardy is that one may take $\theta = 1/2$, and it is known that one cannot do better. For instance, H. Cramér [6] showed that the second moment of the remainder is

$$\frac{1}{T} \int_0^T \left| \frac{N(t) - \pi t^2}{\sqrt{t}} \right|^2 dt = \text{const} > 0$$

and thus we certainly cannot have $\theta < 1/2$.

The distribution of the remainder. In the 1990's, Heath-Brown considered an interesting twist on the problem, by asking for the distribution function of the normalized remainder term

$$S(t) := \frac{N(t) - \pi t^2}{\sqrt{t}},$$

that is, to find a probability distribution $P(s)$ on the line so that the asymptotic fraction of t's such that $a < S(t) < b$ is $\int_a^b P(s)ds$, precisely

$$\lim_{T \to \infty} \frac{1}{T} \text{meas}\{T < t \le 2T : a < S(t) < b\} = \int_a^b P(s)ds$$

where "meas" is Lebesgue measure.

Heath-Brown [8] showed that the limiting distribution $P(s)$ exists, and remarkably it is not the Gaussian distribution. For instance it is skew (it has positive third moment) and has shorter tails than a Gaussian: as $|s| \to \infty$, $P(s)$ decays roughly as $\exp(-s^4)$.

Other integrable systems. The work of Heath-Brown has had a large impact in the mathematical physics community. Bleher, Dyson, Lebowitz, Sinai and others have studied the distribution of the remainder term in Weyl's law for the spectral function $n(x)$ for a number of other "integrable" systems, for instance surfaces of revolution, Liouville tori, as well as other lattice point problems. In all these integrable cases they found a similar feature: the remainder term, properly normalized, has a non-Gaussian limiting distribution. We refer to Bleher's survey [3] for more details and references.

Lattice points in thin annuli. One may also ask for the statistics of the window count $n(x, \delta)$. In the case of the flat torus, this is non other than the number $N(t, \rho)$ of lattice points in an annulus of width $\rho = \delta/2\pi$ and in-radius $t = x/2\pi$. Assuming that t is chosen at random in the interval $[T, 2T]$, we may make the width $\rho = \rho(T)$ vary as well. Here we focus on the "mesoscopic", or "intermediate" regime, where $\rho(T) \to 0$ as $T \to \infty$, so that we have a "thin" annulus, but nonetheless we still insist that the area $2\pi t\rho + \pi\rho^2$ of the annulus (which is the expected number of lattice points) grows to infinity, that is that $T\rho(T) \to \infty$. This problem was studied by Bleher and Lebowitz, who showed [5] that the variance of $N(t, \rho)$ in this regime is

$$\frac{1}{T} \int_T^{2T} |N(t, \rho) - 2\pi t\rho|^2 \sim \text{const} \cdot T\rho |\log \rho|, \quad T \to \infty.$$

The Gaussian conjecture of Bleher and Lebowitz [4] is that the normalized remainder term

$$S(t, \rho) := \frac{N(t, \rho) - 2\pi t\rho}{\sqrt{16t\rho |\log \rho|}}$$

has a standard Gaussian value distribution, in contrast to the non-Gaussian distribution $P(s)$ for $N(t)$, that is

$$\lim_{T \to \infty} \frac{1}{T} \text{meas}\{t \in [T, 2T] : a < S(t, \rho) < b\} = \int_a^b e^{-s^2/2} \frac{ds}{\sqrt{2\pi}}.$$

Results. In 2004, Hughes and the author [9] proved this conjecture for the case that the width of the annulus shrinks "slowly" to zero, that is when $\rho(T) > 1/T^\epsilon$ for all $\epsilon > 0$ (for instance when $\rho(T) = 1/\log T$).

Subsequently, Wigman [12, 13] considered the case of elliptical annuli and showed Gaussian distribution for the case when the aspect ratio is transcendental and strongly Diophantine in a suitable sense, still in the regime where the width is slowly shrinking.

It is an open problem to establish the Gaussian law throughout the mesoscopic regime.

The Berry-Tabor conjecture. A different regime, sometime called the "microscopic" regime, is when the width of the annulus shrinks rapidly, while keeping the area λ of the annulus finite (and nonzero). Let us now assume that we are dealing with a generic elliptical annulus. A special case of a conjecture of Berry and Tabor [1] is that the number $N(t, \rho)$ of lattice points in such an annulus has a Poisson distribution with parameter λ, that is for any integer $k \geq 0$,

$$\lim_{T \to \infty} \frac{1}{T} \text{meas}\{t \in [T, 2T] : N(t, \rho) = k\} = e^{-\lambda} \frac{\lambda^k}{k!}.$$

We cannot expect an interesting distribution for the case of *circular* annuli, as most such annuli will have no lattice points.

The Berry-Tabor conjecture is wide open. The only knowledge we have is that the second moment of $N(t, \rho)$ is as predicted. This was proved by Sarnak [11] for random ellipses, and by Eskin, Margulis and Mozes [7] for Diophantine ellipses.

Chaotic systems. As we saw, the question of the statistics of the spectral counting functions for the flat torus leads to interesting lattice point problems which are still largely unsolved.

The torus falls into the family of surfaces with *integrable* geodesic flow. The other extreme case is that of "chaotic" geodesic flow, for instance negatively curved surfaces. In that case, it is expected that the remainder term for the spectral function $n(x)$ has *Gaussian* fluctuations. Little is known about this, save for the special example of the modular surface [10]. In the microscopic regime, it is believed that for "generic" surfaces the statistics of the spectral counting function $n(x, \delta)$ is modelled by that of suitable random matrix ensembles [2], rather than by the Poisson model. While there have been several investigations of this by physicists and plenty of numerical evidence, virtually nothing is known here in a rigorous fashion.

ACKNOWLEDGEMENTS. I thank the Scuola Normale Superiore of Pisa for its hospitality.

References

[1] M. V. BERRY and M. TABOR, *Level clustering in the regular spectrum*, Proc. Roy. Soc. A **356** (1977), 375–394.

[2] O. BOHIGAS, M. J. GIANNONI and C. SCHMIT, In: "Quantum Chaos and Statistical Nuclear Physics", T. H. Seligman, H. Nishioka (eds.), Lecture Notes in Physics, Vol. 263, Springer-Verlag, Berlin, 1986, 18.

[3] P. BLEHER, *Trace formula for quantum integrable systems, lattice-point problems and small divisors*, In: "Emerging Applications of Number Theory", D. A. Hejhal, J. Friedman, M. C. Gutzwiller, A. M. Odlyzko (eds.), Springer, 1999, 1–38.

[4] P. BLEHER and J. LEBOWITZ, *Energy-level statistics of model quantum systems: universality and scaling in a lattice-point problem*, J. Statist. Phys. **74** (1994), 167–217.

[5] P. BLEHER and J. LEBOWITZ, *Variance of number of lattice points in random narrow elliptic strip*, Ann. Inst. H. Poincaré Probab. Statist. **31** (1995), 27–58.

[6] H. CRAMÉR, *Über zwei Sätze des Herrn G. H. Hardy*, Math. Z. **15** (1922), 201–210.

[7] A. ESKIN, G. MARGULIS and S. MOZES, *Quadratic forms of signature (2, 2) and eigenvalue spacings on rectangular 2-tori*, Ann. of Math. (2) **161** (2005), 679–725.

[8] D. R. HEATH-BROWN, *The distribution and moments of the error term in the Dirichlet divisor problem*, Acta Arithmetica **60** (1992), 389–415.

[9] C. P. HUGHES and Z. RUDNICK, *On the distribution of lattice points in thin annuli*, IMRN **13** (2004), 637–658.

[10] Z. RUDNICK, *A central limit theorem for the spectrum of the modular group*, Ann. Inst. H. Poincaré **6** (2005), 683–883.

[11] P. SARNAK, *Values at integers of binary quadratic forms*, In: "Harmonic Analysis and Number Theory" (Montreal, 1996), CMS Conf. Proc. **21** (1997), 181–203.

[12] I. WIGMAN, *The distribution of lattice points in elliptic annuli*, Q. J. Math., to appear.

[13] I. WIGMAN, *Statistics of lattice points in thin annuli for generic lattices*, Doc. Math. **11** (2006), 1–23.

Algebraic dynamics

Lucien Szpiro and Thomas J. Tucker

Dynamics studies the iterations of a map $\varphi : X \to X$ from a variety X to itself. We will be interested in the case when X is a algebraic variety (or a scheme) and the map is a finite morphism from X to X. In most cases we want the map to be of degree $q > 1$ (this excludes projective varieties of general type, which have no self maps aside from automorphisms). The study of dynamics for complex analytic varieties is classical. Algebraic dynamics has one more tool at its disposal: the canonical height that can be attached to a map $\varphi : X \to X$ when X is a projective variety, X and φ are defined over a number field K, and when, moreover there is an ample line bundle L on X that satisfies

$$\varphi^*(L) = L^d.$$

General properties of such "polarized algebraic varieties" are developed in the papers of S. Zhang [20] and ourselves [13], and in the book of Everest and Ward [6]. Essentially this situation arises when one has (n+1) homogeneous polynomials $(\varphi_0, \varphi_1, \ldots \varphi_n)$, of degree d in (n+1) variables defining a map

$$\varphi : U \to \mathbb{P}^n$$

from the the complement U in $\mathbb{P}^n$ of the scheme of zeroes of the (φ_i), and when X is a closed subscheme of $\mathbb{P}^n$, contained in U and sent to itself by φ. We show in [13] that in this situation $q = \deg(\varphi) = d^{\dim X}$ and that $q > 1$ implies that the set of fixed points of any iterate φ^k is finite.

For example when φ is a map from $\mathbb{P}^1$ to itself the condition on the existence of the line bundle L with the "eigen" property, is automatically verified by the tautological line bundle of $\mathbb{P}^1$. A smooth, projective curve must have genus zero or one if it has an endomorphism of degree at least 2. In fact genus one can be "reduced" to genus zero the following way: muliplication by an integer on an elliptic curve E (as well as complex multiplication) commutes with $[-1]$ and hence can be studied on the quotient of E by $[-1]$, which is the Riemann sphere $\mathbb{P}^1$. For example, if

$$y^2 = P(x)$$

is the Weierstrass equation of E, the multiplication-by-2 map on E gives the following transformation of the x-coordinate:

$$\varphi(x) = \frac{P'(x)^2 - 8x P(x)}{4 P(x)}.$$

Many unsolved problems in Diophantine Geometry are stated as questions about the heights of points of algebraic varieties. We will explain these problems in the text below as they present themselves naturally in term of "canonical heights". Usually one looks at the naive or natural height on projective space generalizing the function log sup $|coordinates|$, for a point with coprime coordinates in $\mathbb{Z}$, that is

$$h(x_0, \ldots, x_n) = \frac{1}{\mathbb{Q}(x_0, \ldots, x_n) : \mathbb{Q}} \log \prod_{\substack{\text{all places}}} \sup |x_i|_v.$$

Siverman and Call [10] have noticed that Tate's proof of the existence of the "Neron-Tate" height on abelian varieties can be exactly copied to define a **canonical height** attached to any (X, φ, L) as above by the following formula:

$$h_\varphi(P) = \lim_{n \to \infty} \frac{h_L(\varphi^n(P))}{d^n}$$

where h_L is any height attached to the line bundle L with and integral model over the ring of integers $\mathcal{O}_K$ and a metrization at places at infinity. The canonical height is characterized by the following properties:

(i) $h_\varphi(\varphi(P)) = d h_\varphi(P)$
(ii) $|h_L - h_\varphi|$ is bounded on $X(\bar{\mathbb{Q}})$

As a consequence the canonical height satisfies the finitness theorem of Northcott:

Points in $\bar{\mathbb{Q}}$ of bounded height and bounded degree make a finite set.

h_φ is also non negative. Moreover the points with canonical height zero are exactly the **preperiodic points** of φ.

One can show that that the canonical height is an Arakelov type height associated associated to adelic metrics (cf. [19]). In particular the curvature of such metrics is the invariant measure associated to φ (see [1]).

The first example is the natural height on projective space described above. It is the canonical height for the map $(x_0, \ldots, x_n) \to (x_0^2, \ldots, x_n^2)$. For $\mathbb{P}^1$ the preperiodic points are are 0, ∞ and the roots of unity.

The second example is the map $\varphi(x) = \frac{P'(x)^2 - 8xP(x)}{4P(x)}$ associated to the multiplication-by-2 in an elliptic curve whose Weirstrass equation is $y^2 = P(x)$. The preperiodic points are the images in $\mathbb{P}^1$ of the torsion points of the elliptic curve.

A third example is given geometrically by the projection of a conic to a line from a point outside the conic. Choosing the conic with equation $XY + pX^2$, for p a prime number, and choosing the projection appropriately leads to the map $\varphi(x) = \frac{x^2+p}{x^2}$. One can note that this map is well defined mod p – it is the projection of a degenerate conic (2 lines).

In fact reducing a dynamical system mod a prime p is an interesting arithmetical operation on the geometry of the situation. This introduce the first notion of **bad reduction**. This intuitive notion is based on the choice of an integral model of $\mathbb{P}^1_K$ over the ring of integers $\mathcal{O}_K$ of the number field K. Then writing our map (possibly after a base change) $\varphi(x) = P/Q$ where P and Q are polynomials with integral coprime coefficients and no common root over the algebraic closure $\overline{K}$ of K one can reduce mod any prime ideal of $\mathcal{O}_K$. The prime $\wp$ will be of bad reduction for this model if P and Q acquire a common root over the algebraic closure of the residue field $k(\wp)$. For example a polynomial map $\varphi(x) = P$ will have bad reduction exactly at primes dividing the leading coefficient of P. In particular a dynamical system given by a monic polynomial will have everywhere good reduction. One can check that the dynamical system on the sphere attached to the multiplication-by-2 of an elliptic curve in Example 2, has bad reduction exactly where the Weierstrass model of the elliptic curve has bad reduction in the theory of elliptic curves (P and P' get a common root mod $\wp$). One can also see that bad reduction is the locus on Spec $\mathcal{O}_K$ where the Zariski closure of the graph of φ in $\mathbb{P}^1_{\mathcal{O}_K} \times \mathbb{P}^1_{\mathcal{O}_K}$ does not project isomorphically on the first factor.

Attached to a model of a dynamical system is **the blowing up of X of the locus in $\mathbb{P}^1_{\mathcal{O}_K}$ where the homogenized P and Q have a common zero.** For example the degenerate conic $XY = 0$ in Example 3 was the fiber of such a blow up. This is a general fact: the map φ can be extended to a finite map of degree $d : X \rightarrow \mathbb{P}^1_{\mathcal{O}_K}$. In the work of J. Pineiro, L. Szpiro and T. Tucker [14] such extension of φ is used to established a **generalized Mahler formula** for the canonical height. If F the minimal equation for P, over $\mathbb{Z}$, the usual Mahler formula for the natural height is:

$$\deg(F)h(P) = \int_0^1 \log|F(\exp(2i\pi\theta))|d\theta.$$

In fact one can verify that $d\theta$ is the invariant measure for the map $(x \to x^2)$ and that the domain of integration-the unit circle- is the closure of the Julia set of this map (the Julia set is the closure of the set of roots of unity *i.e.* the set of repulsive preperiodic points). In the paper quoted above we established an adelic formula with the same features but also possibly with non zero v-adic terms at places of bad reduction in the sense explained above:

$$\deg(F)h(P) = \sum_v \int_{\mathbb{P}^1_{C_v}} \log |F|_v d\mu_v(\varphi).$$

The Mahler formula is one of the few global formula for the height of a point. It established a first relation between the canonical height and the bad reduction. Another result of L. Szpiro and T. Tucker (to appear in the Bogomolov Festchrift in 2006 [15]) relates the bad reduction to bounds of dynamical systems. We have to introduce a more refined notion of bad reduction: the **critical bad reduction**. This is the reduced image in $\mathrm{Spec}(\mathcal{O}_K)$ of the ramified locus of the Zariski closure in $\mathbb{P}^1_{\mathcal{O}_K}$ of the image $\varphi(C)$ of the critical points C of φ. This notion of bad reduction is closely related to the previous one. In good cases it contains the bad reduction defined above. The theorem that we have been able to prove is of the "Shafarevich conjecture type" [15]:

If the cardinal of $\varphi(C)$ is at least 3 and if the degree and the critical bad reduction of φ are fixed, then the set of maps $\varphi(x)$ up to $SL_2 \times SL_2$ equivalence is a finite set.

The change of variables giving the action by of $SL_2 \times SL_2$ on the set of rational maps of degree d has been studied in details by J. Silvermann in his paper "Space of rational maps on $\mathbb{P}^1$" ([9]).

One can develop a notion of **conductor** for a dynamical system defined by a map φ along the lines of the paper of J. Pesenti and L. Szpiro [8]. These authors have defined a "conducteur efficace" (efficient conductor) which can be extended to the situation of a dynamic on the sphere: it is the difference in $\mathrm{Spec}(\mathcal{O}_K)$ between the direct image of the relative differencials of X and the direct image of the relative dualizing sheaf of X. This is the context where the abc conjecture can be generalized: **can the height of the map φ be bounded linearly in term of the log of the norm of the conductor?** This formulation agrees whith the abc conjecture for elliptic curves. The abc conjecture for elliptic curves ask for a bound on the height of the 2-division points. These points are exactly the images of the critical points of φ for $\varphi = \frac{P'(x)^2 - 8xP(x)}{4P(x)}$ where $y^2 = P(x)$ is the Weierstrass equation of the elliptic curve. Bounding the height of

φ or the height of the image of its critical points are equivalent problems. In particular for the example above it would be enough to get a bound for the difference between the canonical height and the natural height.

Some progress on this last problem has been made in the recent work of L. Szpiro and T. Tucker [16] on equidistribution. In the spirit of the paper of L. Szpiro, E. Ullmo and S. Zhang "Equirepatition des petits points" [11], following the work of Bilu [4], Chambert-Loir [5], Bost and Autissier [1] one can prove that the measures involved in the generalized Mahler formula can be computed aginst continuous functions by equidistribution of the Galois Orbits of a sequence of small points (a sequence of small points is a Zariski dense sequence of points with canonical height tending to 0). The important technical theorem use in this context is the Hilbert-Samuel theorem for metrized line bundles (see [12]). One of the more general statement can be found in a recent paper of [18] has been.

The generalized Mahler formula asks for the integration of non continous functions. In the paper "Equidistribution of generalized Mahler measures" [16], using Roth theorem on diophantine approximation we have been able to established that the preperiodic points have their Galois orbits equidistributed for the measures $d\mu_{v,(\varphi)}$ calculated again the log of algebraic functions. R. Rumely and P. Autissier has been able to established a conterexample if instead of a sequence of points of height zero one takes a sequence of small points to calculate

$$\int_{\mathbb{P}^1_{\mathbb{C}_v}} \log |F|_v d\mu_v(\varphi).$$

One corollary of such a theorem is a first result in the evaluation of the difference between canonical and natural height:

$$h_\varphi(P) \leq h(P) + C_\varphi + \log 2$$

where $C_\varphi = \limsup_{Q \text{preperiodic}} h(Q)$. One suspects that an inequality in the other direction has to be also true. In which case the difference between the canonical height and the natural height would be obtained on preperiodic points. More generally one research point of view should be to compare the canonical height associated to 2 different maps φ and ψ but evaluating each on the preperiodic points of the other. For the problems of the type **Lehmer'conjecture** *i.e.* finding lower bounds for the canonical height of non preperiodic points the study of **variations of invariant measures and of Julia sets in families of maps** φ shall shed new light on these classical problems (*cf.* the recent work of Dvornicich and Zannier [17]).

Another direction is the study of **dynamics over a function field**. With
M. Tepper and T. Tucker we are quite advanced in proving the following
statement:

Theorem 1.1 (Conjectural). *Let C be a projective, smooth connected
curve over a field k. Let K be the function field of C. Let X_K be a scheme
of finite type over K with a K endomorphism Φ_K and an ample invertible
sheaf L_K with $\Phi_K^*(L_K) = L_K^{\otimes d}$ for an integer d > 1. Then the set of
points of $X_K(K)$ whose Φ canonical height vanishes, is contained in a
finite number of irreductible preperiodic K-subschemes of X_K which are,
up to a finite radicial morphism, isotrivial with an iterate of the restriction
of Φ_K isotrivial. Moreover when X_K and Φ are trivial the set of points
of canonical height zero is exactly the points defined over k.*

When the variety X is the projective line this theorem was proved by
R. Benedetto and M. Baker (see [3], above [2]). The case of the dynamics
associated to the multiplication by an integer on an elliptic curve (Example 2) is solved in Hindry and Silverman's book [7]. H. Hruchovski has
informed the authors that he has a proof of the statement when there is
no φ – stable subvariety of X_K.

ACKNOWLEDGEMENTS. This survey was written in Pisa when we were
guest of the Centro di Giorgi. We would like to thanks U. Zannier for
numerous interesting mathematical discussions and for his very kind hospitality.

References

[1] P. AUTISSIER, *Points entiers sur les surfaces arithmétiques*, J.
    Reine. Angew. Math. **531** (2001), 201–235.

[2] M. BAKER, *A finiteness theorem for canonical heights attached to
    rational maps over function fields*, arxiv.org/math (2005).

[3] R. BENEDETTO, *A finiteness theorem for canonical heights at-
    tached to polynomial maps over function fields*, preprint (2006),
    1–9.

[4] Y. BILU, *Limit distribution of small points on algebraic tori*, Duke
    Math. **89-3** (1997), 465–476.

[5] A. CHAMBERT-LOIR, *Points de petite hauteur sur les variété semi
    abéliennes*, Annales de l'Ecole Normale Supérieure **33-6** (2000),
    801–821.

[6] G. EVEREST and T. WARD, "Heights of Polynomials and Entropy
    in Algebraic Dynamics", Springer Universitext, 1999, 1–211.

[7] M. HINDRY and J. H. SILVERMAN, *Diophantine geometry an introduction*, Graduate Text in Mathemetics **201** (2000), 1–558.

[8] J. PESENTI and L. SZPIRO, *Inegalite du discriminant pour les surfaces elliptiques a réduction quelconque*, Compositio Math. **120** (2000), 83–117.

[9] J. H. SILVERMAN, *The space of rational maps on* $\mathbb{P}^1$, Duke Math. J. **94-1** (1998), 41–77.

[10] J. H. SILVERMAN and G. S. CALL, *Canonical heights on varieties with morphism*, Compositio Math. **89** (1993), 163–205.

[11] L. SZPIRO, E. ULLMO and S. ZHANG, *Equirépartition des petits points*, Invent. Math. **127** (1997), 337–347.

[12] L. SZPIRO, *Algebraic Geometry over* $\overline{\mathbb{Q}}$, London Mathematical Society Lecture Notes Series **238** (1997), 117–123.

[13] L. SZPIRO and T. J. TUCKER, *Ample polarization*, (2006).

[14] J. PINEIRO, L. SZPIRO and T. J. TUCKER, *Malher measure for dynamical systems on* $\mathbb{P}^1$ *and intersection theory on a singular arithmetic surface*, Progress in Math: Geometric Methods in Algebra and Number Theory **235** (2004), 219–250.

[15] L. SZPIRO and T. J. TUCKER, *A Shafarevich-Faltings theorem for rational functions*, In: "Quarterly Journal of Pure and Applied Mathematics", for the occasion of the sixtieth birthday of Fedor Bogomolov., Volume in honor of F. Bogomolov.

[16] L. SZPIRO and T. J. TUCKER, *Equidistribution and generalized Mahler measures*, preprint to appear (2006), 1–30.

[17] R. DVORINICICH and U. ZANNIER, *Cyclotomic diophantine problems*, preprint Scuola Normale Superiore di Pisa, 2006.

[18] X. YUAN, *Big line bundles over arithmetic varieties*, preprint Columbia University, 2006.

[19] S. ZHANG, *Small points and adelic metrics*, Journal of Algebraic Geometry **4** (1995), 281–300.

[20] S. ZHANG, *Distribution and heights in algebraic dynamics*, preprint (2006), 1–60.

COLLOQUIA

The volumes of this series reflect lectures held at the "Colloquio De Giorgi" which regularly takes place at the Scuola Normale Superiore in Pisa. The Colloquia address a general mathematical audience, particularly attracting advanced undergraduate and graduate students.

Fotocomposizione "CompoMat" Loc. Braccone, 02040 Configni (RI) Italy
Finito di stampare nel mese di aprile 2007
dalla CSR srl, Via di Pietralata, 157, 00158 Roma